Christfried Jakob, Joseph Collins

An Atlas of the Normal and Pathological Nervous Systems

Christfried Jakob, Joseph Collins

An Atlas of the Normal and Pathological Nervous Systems

ISBN/EAN: 9783337408015

Printed in Europe, USA, Canada, Australia, Japan

Cover: Foto ©berggeist007 / pixelio.de

More available books at **www.hansebooks.com**

AN ATLAS

OF THE

NORMAL AND PATHOLOGICAL

NERVOUS SYSTEMS.

TOGETHER WITH A SKETCH

OF THE

ANATOMY, PATHOLOGY, AND THERAPY OF THE SAME.

BY

DR. CHRISTFRIED JAKOB.

PRACTISING PHYSICIAN IN BAMBERG,
FORMERLY FIRST ASSISTANT IN THE MEDICAL CLINIC AT ERLANGEN.

WITH AN INTRODUCTION BY

PROF. DR. AD. v. STRÜMPELL.

Translated and Edited (authorized) by

JOSEPH COLLINS, M.D.,

INSTRUCTOR OF NERVOUS AND MENTAL DISEASES, NEW YORK POST-GRADUATE MEDICAL SCHOOL;
VISITING PHYSICIAN TO THE HOSPITAL FOR NERVOUS DISEASES;
ATTENDING PHYSICIAN TO ST. MARK'S HOSPITAL.

NEW YORK:

WILLIAM WOOD & COMPANY,

1896.

PREFATORY NOTE.

DURING his tenure of assistantship in the medical clinic of this place Dr. Jakob has busied himself in a most industrious manner with the normal and pathological anatomy of the nervous system. By means of most approved methods of investigation he has prepared with great care a large collection of histological preparations, and these in connection with the uncommon talent for drawing which he possesses have put him in a position to prepare the following atlas almost entirely from his own preparations and drawings. Every unprejudiced observer will, like myself, I think, be convinced that the illustrations convey everything that can be thus given. They give the actual relationships in a clear and lucid manner, and they depict with great completeness nearly all the numerous important discoveries which the last decades have brought forth. The student as well as the practical physician who wishes to keep in touch with the advance which medical science has made from the standpoint of the neurologist is given the opportunity to get a clear conception of it by means of this atlas. The intimate relationship existing between clinical pathology on the one hand and normal and pathological anatomy on the other is scarcely so strik-

PREFATORY NOTE.

ing or apparent in any other branch of medicine as in neuropathology.

The individual study of the normal anatomical and pathological conditions in connection with the illustrations which are detailed but not schematic, true representations of the parts as they exist, cannot help but be of the greatest didactic use.

I wish heartily for this work, which the author and publisher have spared no pains to make really good and useful, the success which it deserves.

DR. STRÜMPELL.

AUTHOR'S PREFACE.

IN the present atlas I have endeavored to portray an important chapter of medicine, which usually remains, for the majority of students and physicians of all classes, foreign and unenjoyable. I have, moreover, endeavored to take this bugbear and to place its usual normal and pathological anatomy in comprehensible form. It has been my idea to make easy of comprehension for students the clinical picture as seen at the bedside and the processes on which it depends; and at the same time to make it possible for the practical physician, who has heretofore stood aloof from this subject, to understand the meaning of the most important recently developed facts of neurology. To this end I have made the freest use of illustrations. In the text everything unnecessary, especially of a histological nature, I have relegated as much as possible to the background.

The illustrations serve substantially to show the facts as they actually exist, without being schematic. The exhibition of stained specimens was unavoidable. Unstained fresh specimens have, however, been used. The actual study of the fresh brain can be substituted by no form of illustration, even photographic. The reproduction by lithographs and woodcuts has been

AUTHOR'S PREFACE.

carried out with unusual care under my entire control.

The source of the illustrations is the pathological collection which I have been able to arrange during my assistanceship of several years in the medical clinic at Erlangen from its very rich material. I owe the greatest thanks to my former respected chief, Prof. Dr. Strümpell, for the liberality with which he has always placed everything necessary at my disposal, for the gracious support which he has always given me in every question, and in this place I may be permitted to make acknowledgment of the same.

A few words on the study of the illustrations: From the natural limitations of the pages devoted to text it has been impossible to describe in detail the wealth of facts which are to be seen in the figures. The pictures, therefore, contain decidedly more material than is explained.

I consider this indeed rather an advantage for the book. I hope that by this means actual close study will be directed toward the illustrations. And thus the formation of an independently reached decision of the greatest importance for our science may be made possible. The material presented appears to me in a certain way to tend in that direction.

CHRISTFR. JAKOB.

BAMBERG, Summer, 1895.

TRANSLATOR'S PREFACE.

At the request of the publishers I have undertaken the translation of Jakob's explanatory atlas of the normal and diseased histology of the nervous system because of my conviction that the volume would be of assistance to students and to the general practitioner. Feeling it a duty to the author, I have endeavored to make the translation a literal one so far as it was consistent with lucidity. This has been done in some instances at the expense of style and smoothness of diction.

A word of explanation is necessary in reference to the translation of certain words. Gehirnstamm, a word proposed by Meynert to include that part of the brain left after removal of the hemispheres and basal ganglia, has been rendered by the word brain stem as a rule, exceptionally by the word brain axis. Anlage, an extremely useful but untranslatable word which by some authors has been taken over bodily into English, has been translated in most instances by the word proton, a term suggested by Prof. B. G. Wilder, and the paronymization protal—two euphonious, significant words. Occasionally it has been rendered "primary constituent," especially when used in biological connection. Such descriptive words as

TRANSLATOR'S PREFACE.

"Faserendaufsplitterung" have been rendered literally terminal arborization of fibres.

I have also taken the liberty to use the terms ventral and dorsal for anterior and posterior in some instances, and likewise the now almost universally used terms caudal and cephalic for spinalwärts and cerebralwärts. Editorial additions or comments are in brackets.

JOSEPH COLLINS.

NEW YORK, October 15, 1895.

REGISTER OF THE PLATE ILLUSTRATIONS.

SPECIFICATION OF THE MAGNIFICATION, STAINING, ORIGIN, ETC., OF THE PREPARATIONS.

1. Morphology of the Central Nervous System.

PLATE 1.—Brain *in situ* looked at from above after removal of the bony covering. Revealed by turning the first layer of pasteboard back.

Schematic original drawing ⅔ natural size. Enlarged from the plan of a drawing by Oestricher.

PLATE 2.—Horizontal cross section of the cerebrum at the level of the corpus callosum after removal of the latter. The fornix to be turned back.

Original drawing ⅔ natural size from a fresh brain.

PLATE 3.—Horizontal cross section of the brain after laying open the third ventricle. On the right the superfices of the central ganglia cut through horizontally.

Original drawing ⅔ natural size from a fresh brain.

PLATE 4.—Horizontal cross section of the brain through the central ganglia (on the left a higher section, on the right a deeper). The cerebellum is uncoverable.

Original drawing ⅔ natural size from a fresh brain.

PLATE 5.—Base of the brain with nerves and blood-vessels.

Original drawing ⅔ natural size from a fresh brain.

PLATE 6.—Fig. 1. Base of the cerebrum after removal of brain stem and cerebellum.

Schematic original drawing based on a drawing from Eberstaller.

LIST OF PLATES.

Fig. 2. Survey of the projection tracts.
Schematic original drawing.

PLATE 7.—Four frontal sections through the brain of a dog.
Fig. 1. Through the frontal lobes.
Fig. 2. Through the optic thalamus.
Fig. 3. Through the cerebral peduncle.
Fig. 4. Through the occipital lobes.
Original photographs from fresh cross-sections.

PLATE 8.—Frontal sections through the human brain stem.
Fig. 1. Section through the middle of the third ventricle.
Fig. 2. Through the mammillary bodies.
Fig. 3. Through the posterior end of the third ventricle.
Three original photographs from preparations. Natural size.

PLATE 9.—Ten frontal sections through the brain stem; parallel cross sections of the medulla and spinal cord.
Fig. 1. Anterior corpora quadrigemina.
Fig. 2. Middle of pons.
Fig. 3. Middle of fourth ventricle.
Fig. 4. Section below 3.
Fig. 5. Posterior end of the fourth ventricle.
Fig. 6. Medulla oblongata below the fourth ventricle.
Fig. 7. Pyramidal decussation.
Fig. 8. Cross section of cervical cord.
Fig. 9. Cross section of dorsal cord.
Fig. 10. Cross section of lumbar cord.
Original photographs from author's preparations. Natural size.

PLATE 10.—Cross section of spinal cord *in situ*.
Fig. 1. Cervical cord of a child, with its roots in the spinal canal.
Original photograph from personal preparation. Magnified 2½ diameters. Medullary sheath staining.

Fig. 2. Lumbar cord and membranes of a new-born.
Original photograph from personal preparation. Magnified 10 diameters.

LIST OF PLATES.

2. Development and Structure of the Nervous System.

PLATE 11.—Embryonal primary constituent.
 Fig. 1. Cross section through the medullary grooves (schematic).
 Fig. 2. Cross section through the medullary channel.
 Fig. 3. Cross section of the spinal cord at second month (carmine stain).
 Fig. 4. Cross section of spinal cord at eighth month (medullary sheath stain).
 Fig. 5 Peripheral nerve of a new-born (medullary sheath stain).
> Original drawings (Fig. 1 with the use of an illustration from Merkel. Fig. 2, one from His) from personal preparations.

PLATE 12.—Embryonal development of the brain.
 Fig. 1. Brain and spinal cord, third month.
 Fig. 2. Base of brain, fourth month.
 Fig. 3. Median surface of brain, fourth month.
 Fig. 4. External surface of the cerebral hemisphere, fifth month.
> Original drawing. Natural size.

PLATE 13.—Structure of neurons.
 Fig. 1. Pyramidal cells (metallic impregnation after Golgi).
 Fig. 2. Anterior horn cells (stain as above). Both magnified 700 diameters.
 Fig. 3. Scheme of the construction of the more important tracts and their connections.
> Original drawings.

PLATE 14.—Ganglion cells.
 Fig. 1. *a–d*, original drawings from preparations with Nissl's stain. × 400.
 Fig. 2. Purkinje's cells (Cajal's method).
> Original photographs. × 250.

PLATE 15.—Cerebral cortex.
 Fig. 1. Cortex of frontal lobe.
 Fig. 2. Cortex of occipital lobe.
> Original drawings based on an illustration from Edinger. On the left methyl blue stain, on the right medullary sheath stain.

LIST OF PLATES.

PLATE 16.—Brain cortex and meninges.
 Fig. 1. Scheme of the membranes.
 Fig. 2. Choroid plexus. × 150, carmine staining.
 Fig. 3. Photograph of a cerebral convolution (medullary sheath stain). × 50 diameters.
 Fig. 4. Cerebellar cortex (combined medullary sheath and methyl blue staining). × 60 diameters.
  Original drawings, personal preparations.

PLATE 17.—Cerebral convolutions.
 Figs. 1 and 2. Schematic representation of the convolutions on the external and median surface of the hemispheres.
  Original drawings.

 Fig. 3. The cortical centres and their location in the skull cavity.
  Original drawing based on a figure from Vierordt.

PLATE 18.—Fig. 1. The position of the motor and sensory nuclei in the brain stem and medulla.
  Schematic original drawings.

 Fig. 2. *a–d*, nerve fibre, longitudinal and cross section. *a*, *b*, medullary sheath stain. *c*, freshly isolated. *d*, nigrosin staining.

PLATE 19.—Lateral view of the medulla; representation of the nuclei of the cranial nerves.
  Schematic original drawings based on a representation from Edinger.

PLATE 20.—The nuclear layer of the motorial peripheral neurons.
 Fig. 1. Section through the hypoglossal nucleus.
 Fig. 2. Section through anterior horn of the cervical cord.
  Original drawing from (combined) preparations stained with carmine medullary sheath stain. × 360 diameters.

PLATE 21. Fig. 1. Cross section through the white medullary substance of the spinal cord (lateral columns).
 Fig. 2. Section through a spinal ganglion.
 Fig. 3. Section through a peripheral nerve.
  Original drawings. Fig. 1, carmine preparation; Fig. 2, carmine and medullary-sheath stain; Fig. 3, nigrosin staining. × 360 diameters.

LIST OF PLATES.

PLATE 22.—Fig. 1. Schematic representation of the composition of the cervical and lumbar cord.

Fig. 2. Central canal in lumbar region (medullary sheath stain)

> Original drawings. Fig. 2 × 150 diameters.

3. Topographical Anatomy of the Nervous System. Serial Sections. Schematic.

PLATE 23.—Division of the brain and spinal-cord nerves.

> Original drawing with use in part of illustration from Gowers.

PLATE 24.—Fig. 1. Frontal section through the knee of the corpus callosum and the anterior portion of the frontal lobes.

Fig. 2. Frontal section through the head of the caudate nucleus.

> These and all the following original drawings are from preparations made with medullary sheath stains. Natural size.

PLATE 25.—Fig. 1. Frontal section in the middle of the septum pellucidum.

Fig. 2. Section through the anterior commissure.

PLATE 26.—Fig. 1. Frontal section behind the anterior commissure.

Fig. 2. Section through the knee of the internal capsule.

PLATE 27.—Fig. 1. Section through the middle commissure.

Fig. 2. Section through the central convolutions.

PLATE 28.—Fig. 1. Section through the parietal lobes.

Fig. 2. Section through the parietal lobes.

PLATE 29.—Fig. 1. Frontal section through the posterior pole of the occipital lobes.

Fig. 2. Horizontal section through the superfices of the corpus striatum and thalamus opticus.

Fig. 3. Horizontal section through the cerebral peduncles.

PLATE 30.—Horizontal section through the entire left hemisphere in the middle of the central ganglia.

PLATE 31.—Horizontal section through the base of the brain stem of the left hemisphere.

LIST OF PLATES.

PLATE 32.—Fig. 1. Right angle section through the anterior corpora quadrigemina.

Fig. 2. Section between anterior and posterior corpora quadrigemina.

× 1½ diameters, and the same for the following.

PLATE 33.—Fig. 1. Section through the posterior corpora quadrigemina.

Fig. 2. Section through the middle of the pons.

PLATE 34.—Fig. 1. Section through the posterior end of the pons.

Fig. 2. Section through the acusticus nuclei.

PLATE 35.—Fig. 1. Section through the right optic thalamus at the level of the middle commissure.

Original photographs from medullary-sheath stain preparations. × 1½ diameters. The same for the following.

Fig. 2. Section through the anterior quadrigeminal body of the left side.

PLATE 36.—Fig. 1. Section through the tegmentum behind the posterior quadrigeminal bodies.

Fig. 2. Section through the nuclear region of the trigeminus.

PLATE 37.—Fig. 1. Section through the right tegmental region at the level of the facial nucleus.

Fig. 2. Section through the ventral acoustic nucleus of the left side.

PLATE 38.—Fig. 1. Section through the cerebellum and medulla oblongata (slightly magnified).

Fig. 2. Section through the medulla oblongata at the level of the glosso-pharyngeal-vagus nuclei.

Fig. 2, × 10 diameters.

PLATE 39.—Fig. 1. Section through the medulla, at the level of the tenth and twelfth nuclei.

Fig. 2. Section through the calamus scriptorius of the medulla. × 10 diameters.

Original photographs; the following likewise.

PLATE 40.—Fig. 1. Section through the nuclei of the posterior columns.

Fig. 2. Section through the medulla below the olives.

LIST OF PLATES.

PLATE 41.—Section immediately above the pyramidal decussation.
  Fig. 2. Section through the pyramidal decussation.
PLATE 42.—Fig. 1. Section through the cervical cord immediately below the pyramidal decussation.
  Fig. 2. Section through the upper cervical cord at the level of the fourth cervical nerves.
PLATE 43.—Fig. 1. Section through the cervical enlargement at the level of the seventh cervical nerves.
  Fig. 2. Section through the upper dorsal cord at the level of the third dorsal nerves.
  Fig. 3. Section through the mid-dorsal cord level of sixth dorsal nerves.
PLATE 44.—Fig. 1. Section through the lower dorsal cord level of eleventh dorsal nerves.
  Fig. 2. Section through upper lumbar cord level of second lumbar nerves.
  Fig. 3. Section through lower lumbar cord level of fourth lumbar nerves.
PLATE 45.—Fig. 1. Section through middle sacral cord level of third sacral nerves.
  Fig. 2. Section through the cauda equina and the conus medullaris.
  Fig. 3. Section through the posterior root and a spinal ganglion from the lumbar region.
    Fig. 3, × 20 diameters.
PLATE 46.—Fig. 1. Cross section through the entire sciatic nerve at its upper point of exit.
  Fig. 2. Longitudinal section of a nerve bundle from the sciatic.
  Fig. 3. A nerve bundle from the sciatic on cross section.
  Fig. 4. Cross section from the normal optic nerve.
    Fig. 1, × 10 diameters. Figs. 2 and 3, × 150 diameters. Fig. 4, × 10 diameters.
PLATE 47. The gray substance of the spinal cord.
  Fig. 1. One-half of cervical cord.
  Fig. 2. One-half of lumbar cord.
    Original photographs. Medullary sheath staining. × 25 diameters.

LIST OF PLATES.

PLATE 48.—Medullary sheath proton in the fœtal brain.
  Figs. 1 and 2. Cross section of the thalamus.
  Fig. 3. Anterior corpora quadrigemina.
  Fig. 4. Pons.
  Fig. 5. Medulla.
> Original photographs from unstained preparations. Natural size.

  Fig. 6. Cervical cord.
  Fig. 7. Dorsal cord of a new-born.
> Medullary sheath staining. × 10 diameters.

PLATE 49.—Scheme of the course of the more important brain tracts.
> Original drawings; likewise the following.

PLATE 50.—Figs. 1 and 2. Composition of crusta and tegmentum.
  Fig. 3. Schematic representation of the course of the motor-oculi and optic nerves.

PLATE 51.—Figs. 1 and 2. Scheme of the course of the pyramidal and sensory tracts.
  Fig. 3. Scheme of the sensory nerve tracts.
> Fig. 3 with use of an illustration from Lenhossék.

PLATE 52.—Scheme of the course of the fibres in the spinal cord.
> Original drawing.

4. General Pathological Anatomy of the Nervous System.

Special Pathology of the Brain.

PLATE 53.—Secondary diseases of the nervous system.
  Fig. 1. Section through the brain cortex and meninges in epidemic cerebro-spinal meningitis.
  Fig. 2. Cerebral cortex in tubercular meningitis.
> Original drawings from my own preparations. Cases from the medical clinic at Erlangen. Likewise the following.

LIST OF PLATES.

PLATE 54.—Fig. 1. Aneurism in the region of the quadrigeminal bodies.
Fig. 2. Caries of the body of a vertebra.
Fig. 3. Tumor of the spinal dura.
(Combined medullary sheath-carmine staining.)

PLATE 55.—Fig. 7. Gumma of the base of the brain. Primary diseases of the nervous system.
Fig. 2. Section through the anterior horn of the cervical cord from a case of spinal muscular atrophy.
Fig. 3. Scheme of secondary degeneration in the spinal cord.

PLATE 56.—Disease of the ganglion cells.
PLATE 57.—Disease of the nerve fibres.
PLATE 58.—Disease of the muscle fibres.

All original drawings from personal preparations. Cases in great part from the Erlangen Medical Clinic except Plate 56, Fig. 7, which are after a drawing by Oppenheim.

PLATE 59.—Fig. 1. Porencephaly, left hemisphere.
Fig. 2. Hemorrhagic focus in the brain stem.

Original photographs. Cases from the Erlangen Medical Clinic.

PLATE 60.—Fig. 1. Corpora quadrigemina focus.
Fig. 2. Solitary tubercle in the medulla.
Fig. 3. Chronic progressive ophthalmoplegia.

Original photographs from my preparations. Medullary sheath staining × 2 to 5 diameters. Cases from the Erlangen Medical Clinic.

Secondary Degenerations.

All original photographs, personal preparation. Medullary-sheath staining. Cases from the Erlangen Medical Clinic.

PLATE 61.—Figs. 1 and 2. Total degeneration of the corona radiata to the crusta.
Fig. 3. Degeneration of the frontal pontine tract in the crusta.

PLATE 62.—Figs. 1 and 2. Pyramidal-tract degeneration in the crusta and internal capsule.

LIST OF PLATES.

PLATE 63.—Fig. 1. Pyramidal-tract degeneration in the pons.
Fig. 2. Tegmental-tract degeneration in the frontal tegmentum.

PLATE 64.—Fig. 1. Pyramidal-tract degeneration in the medulla oblongata.
Fig. 2. Degeneration in the medulla in a case of infantile paralysis.
Fig. 3. Fillet degeneration in the medulla oblongata.

PLATE 65.—Descending total degeneration of the pyramidal tract in the spinal cord, cervical, dorsal, lumbar, and sacral, from a focus in the brain.

PLATE 66.—Descending degeneration in the spinal cord from a spinal-cord focus.

> Original preparation, photographs; Cases from which Figs. 1-3 were made have been published by Daxenberger.

PLATE 67.—Ascending degeneration in the spinal cord.

PLATE 68.—Ascending degeneration in the upper cervical cord and the medulla oblongata.

5. Special Pathology of the Spinal Cord and the Peripheral Nerves.

PLATE 69.—Forms of myelitis.
Fig. 1. Column-form, acute (toxic) myelitis, cervical cord.
Fig. 2. Chronic (syphilitic) myelitis, dorsal cord.
Fig. 3. "Compression myelitis."

> Original photographs, personal preparation, medullary sheath staining. × 8 to 10 diameters. All cases, when not otherwise stated, from the Erlangen Medical Clinic.

PLATE 70. Syringomyelia.
Fig. 1. Hydromyelia, upper dorsal cord.
Figs. 2 and 3. Sections through the upper and middle cervical cord from another case of syringomyelia.

> Fig. 1 from one of Professor v. Strümpel's preparations. Carmine staining.

PLATE 71.—Multiple cerebro-spinal sclerosis.
Fig. 1. Longitudinal section of dorsal spinal cord.
Figs. 2 and 4. Cross section through dorsal cord.
Fig. 3. Section through the cerebral cortex.

LIST OF PLATES.

PLATE 72.—Fig. 1. Section through the medulla oblongata from a case of chronic bulbar paralysis with amyotrophic lateral sclerosis.

Fig. 2. Section through the posterior cervical cord in amyotrophic lateral sclerosis.

Fig. 3. Section through the anterior horn of the cervical cord in spinal muscular atrophy (Professor v. Strümpell).

PLATE 73.—Tabes.

Fig. 1. Section through lower dorsal cord.

Fig. 2. Section through upper lumbar cord of another case.

Fig. 3. Section through lower lumbar cord of a third case.

PLATE 74.—Fig. 1. Section through the cervical cord of the case Plate 73, Fig. 1.

Fig. 2. Section through the cervical cord; case of so-called "high tabes."

Fig. 3. Section through the upper cervical cord (tabes combiné).

PLATE 75.—Spastic spinal paralysis.

Fig. 1. Section through the medulla oblongata.

Fig. 2. Section through the cervical cord.

Fig. 3. Section through the middle dorsal cord.

Fig. 4. Section through the lumbar cord. Case of Professor v. Strümpell.

PLATE 76.—Combined systemic diseases.

Fig. 1. Section through the cervical cord.

Fig. 2. Section through the lower dorsal cord.

Fig. 3. Section through the lumbar cord.

PLATE 77.—Degeneration of peripheral nerves.

Fig. 1. Ascending secondary degeneration in the sciatic nerve after amputation of the leg.

Fig. 2. Descending degeneration of the right optic nerve after destruction of the left lateral geniculate body and the right corpora quadrigemina.

Fig. 3. Compression neuritis of the optic nerve.

Fig. 4. Motor nerve branch in spinal muscular atrophy.

Fig. 5. Posterior root from lumbar cord, case of tabes.

Fig. 6. Bundle from the peroneal nerve in neurotic muscular atrophy.

Figs. 1, 4, 5, 6, magnified 350 diameters. Figs. 2 and 3, magnified 10 diameters

LIST OF PLATES.

PLATE 78.—Multiple neuritis.

Figs. 1 and 2. Preparations from a severe alcoholic polyneuritis (1, from crural nerve; 2, from sciatic nerve).

Fig. 5. Transverse section of cervical cord of same case.

Fig. 3. Post-diphtheritic neuritis.

Fig. 4. Longitudinal section through a nerve bundle from the sciatic nerve in infectious polyneuritis (Landry's paralysis).

× 350 diameters. Cases of Figs. 1 and 2 from the Nüremberg Hospital. The others are personal observations.

Apparatus used: Leitz's microscope.
Leitz's microphotographic camera.
Edinger's drawing apparatus and Leitz's camera.

CONTENTS OF THE TEXT.

SECTION I.
Morphology of the Nervous System.
(Plates 1 to 10.)

PAGE

Synopsis of the External Configuration, Construction, and Position.
Meninges, ... 1
Cerebral Hemispheres and Convolutions, . . . 2
Substance of the Hemispheres and Ventricular System, . 5
Thalamus and Third Ventricle, ... 9
Corpora Quadrigemina, ... 12
Cerebellum and Fourth Ventricle, ... 14
Pons and Medulla, ... 15
Spinal Cord, ... 17
Cerebral Nerves, ... 18
Spinal Nerves, ... 19
Sympathetic, ... 20
Blood-vessels, Nerves, ... 21

SECTION II.
Development and Structure of the Nervous System.
(Plates 11 to 48.)

Synopsis of the Ontology and Histology of the Nervous System. Description of the Most Important Tracts.
Proton of the Medullary Canal, ... 23
Proton of the Brain, ... 24
Spongioblasts, Neuroblasts, ... 25

xxi

CONTENTS OF THE TEXT.

	PAGE
Motor and Sensory Roots,	26
Development of the Medullary Sheath,	27
Microscopic Structure of the Glia, Ganglion Cells, Axis Cylinders, and Nerve Fibres,	28
Neuron,	29
The Commissural Tracts,	30
The Association Tracts,	30
The Projection Tracts,	31

SECTION III.

Anatomy and Physiology of the More Important Nerve Tracts.

(Plates 49 to 52.)

1. The Motorial Tract.
   General Remarks, ... 39
   Motorial Cerebral Nerves, ... 41
   Motorial Spinal Nerves, ... 45
2. The Sensory Pathway.
   General Remarks, ... 47
   Course in the Spinal Cord, ... 49
   Course in the Brain, ... 51
   Sensory Cranial Nerves, ... 57
3. The Reflex Tracts and the Reflex Act, . . . 62
4. The Tract of the Will, ... 64
   Cortical Function, Association, ... 65
   Speech and its Tracts, ... 68
5. Tracts of Co-ordination and their Function, . . 72

SECTION IV.

General Pathology and Therapeutics of the Diseases of the Nervous System.

(Plates 53 to 68.)

1. Causation of the Diseases of the Nervous System, . . 75
2. The Pathologico-Anatomical Changes in Nervous Diseases in General, ... 79

CONTENTS OF THE TEXT.

	PAGE
3. General and Special Remarks on the Symptomatology and Localization of Nervous Diseases. Topical Diagnosis,	82
General Remarks,	83
I. Focal Diseases,	85
A. Of the Brain,	85
Convolutions,	86
Medullary Substance,	89
Brain Stem (Corpora Quadrigemina, Thalamus),	91
Pons, Medulla,	93
Cerebellum,	94
Base of the Brain,	94
B. Focal Symptoms of the Spinal Cord,	96
Half-Sided Lesions,	96
Cervical Cord,	97
Dorsal Cord,	98
Lumbar Cord,	99
Sacral Cord,	99
Cauda Equina,	100
C. Symptoms in Lesion of the Peripheral Nerves,	100
Plexus Paralysis,	101
Lesions of the Cranial Nerves,	102
Lesions of the Spinal Nerves,	105
D. Symptoms of Disease of the Sympathetic,	109
II. Symptomatology of the System Diseases,	110
4. General Remarks Concerning the Mode and Method of Examination, Together with a Diagnostic Survey.	
A. History,	113
B. Present Condition,	113
I. Examination of the Motor Sphere,	114
1. Inspection, Mensuration,	114
2. Motorial Irritative Symptoms,	114
3. Motorial Strength,	115
4. Power of Co-ordination,	116
5. Electro-diagnosis,	116
a. Galvanic,	116
b. Faradic Examination,	120

CONTENTS OF THE TEXT.

	PAGE
II. Examination of the Sensory Sphere,	122
1. Subjective Sensations,	122
2. Cutaneous Sensibility (Tactile, Temperature Sense, etc.),	122
3. Sensibility of the Deeper Parts,	125
4. The Higher Sensibility,	126
The Sense of Sight,	126
Sense of Hearing,	127
Sense of Smell, Taste,	128
III. Investigation of the Reflexes,	128
General Remarks,	128
Skin and Tendon Reflexes,	129
Pupillary Reflexes,	132
IV. Investigation of the Function of the Bladder and Rectum,	133
V. Examination for Trophic and Vasomotor Disturbances,	134
VI. Examination of the Psychical Functions,	135
Speech and Writing,	135
Mental Examination,	140
Other Psychical Disturbances,	140
C. The Making of the Diagnosis,	141
5. General Remarks on the Treatment of Nervous Diseases,	142
1. The Prophylaxis,	142
2. Causal Therapy,	143
Treatment by Abstinence,	143
Syphilis, Malaria,	144
Surgical Treatment of	
(a) The Brain,	144
(b) The Spinal Cord,	145
(c) Peripheral Nerves,	146
3. Symptomatic Therapy,	146
Psychical,	146
Physical,	147
The Administration of Drugs,	148

SECTION V.

Special Pathology and Therapy.

(Plates 53 to 78.)

I. Diseases of the Membranes and Blood-Vessels of the Brain,
  1. Pachymeningitis interna hæmorrhagica, . . 150
  2. Acute Leptomeningitis,
      a. Epidemic Cerebro-Spinal Meningitis, . . 151
      b. Purulent Meningitis, ... 152
  3. Tubercular Meningitis, ... 153
  4. Syphilitic Meningitis, Gummatous Meningitis, and Brain Syphilis, ... 154
  5. Thrombosis of the Sinuses, ... 156

II. Diseases of the Brain Substance.
  A. Organic Diseases.
    1. Disturbances of Circulation and its Consequences.
        a. Anæmia and Hyperæmia of the Brain, . 157
        b. Hemorrhage into the Brain, . . . 158
        c. Embolism of the Brain, . . . 162
        d. Aneurismal Formation, ... 163
        e. Arteriosclerosis, ... 163
    2. Inflammatory Diseases of the Brain Substance.
        a. Abscess of the Brain, ... 164
        b. Acute Non-Purulent Encephalitis, . . 165
    3. Tumor of the Brain, ... 166
    4. Internal Hydrocephalus, ... 169
    5. Dementia Paralytica, ... 170
    6. Ophthalmoplegia, ... 172
    7. Bulbar Paralysis, ... 173
    8. Disease of the Cerebellum, ... 175
  B. Brain Diseases of Unknown Nature and Seat (Neuroses of the Brain).
    1. Neurasthenia, ... 176
    2. Hypochondria, ... 177

CONTENTS OF THE TEXT.

		PAGE
3.	Hysteria,	178
4.	The Traumatic Neuroses,	182
5.	Hemicrania,	183
6.	Cephalalgia, Habitual Headache,	183
7.	Genuine Epilepsy,	184
8.	Infantile Eclampsia,	186
9.	Chorea Minor,	186
10.	Chronic Hereditary Chorea, Tic Convulsif,	187
11.	Paralysis Agitans,	188
12.	Myotonia Congenita,	188

III. The Diseases of the Spinal Cord.
  A. Transverse Lesions.
  Diseases of the Membranes of the Spinal Cord, . 188
    1. Hypertrophic Cervical Pachymeningitis, . 189
    2. Syphilitic Spinal Meningitis, . 190
    3. Compression of the Spinal Cord, . 191
    4. Acute and Chronic Myelitis, . 194
    5. Syringomyelia, . 196
    6. Hemorrhages into the Canal of the Spinal Cord, . 197
    7. Multiple Cerebro-Spinal Sclerosis, . 198
  B. System Diseases.
    8. Spastic Spinal Paralysis, . 199
    9. Amyotrophic Lateral Sclerosis, . 200
    10. Spinal Progressive Muscular Atrophy, . 201
        Neural Muscular Atrophy, . 202
    11. Progressive Muscular Dystrophy, . 202
    12. Poliomyelitis Anterior:
        Acute, . 205
        Chronic, . 206
    13. Tabes Dorsalis, . 207
    14. Hereditary Ataxia, . 210

IV. Diseases of the Peripheral Nerves.
  A. Diseases of Individual Nerves.
    Etiology, . 211
    1. Diseases of the Motor Nerves, . 212
        Individual Paralyses, . 213
        Localized Muscular Spasms, . 213

CONTENTS OF THE TEXT.

		PAGE
2. Diseases of Sensory Nerves,	. .	. 214
Neuralgias, ...	. .	. 215
B. Multiple Neuritis.		
Alcoholic, Diphtheritic, etc.,	. .	. 216
Infectious Polyneuritis,	. . .	. 219
Infectious Polymyositis,	. . .	. 219
V. Other Diseases of the Nervous System in Part of Unknown Nature and Seat.		
1. Basedow's Disease, . . .		. 220
2. Myxœdema, ...		221
3. Acromegaly, ...		222
4. Tetany, ...		222
Tetanus, ...		. 223

SECTION VI.

Remarks upon the Conduct of an Autopsy and upon the Methods of Procedure in the Microscopic Investigation of the Nervous System, ... 224

I.

MORPHOLOGY OF THE CENTRAL NERVOUS SYSTEM.

(Section I. of Textual Part.)

Tab.

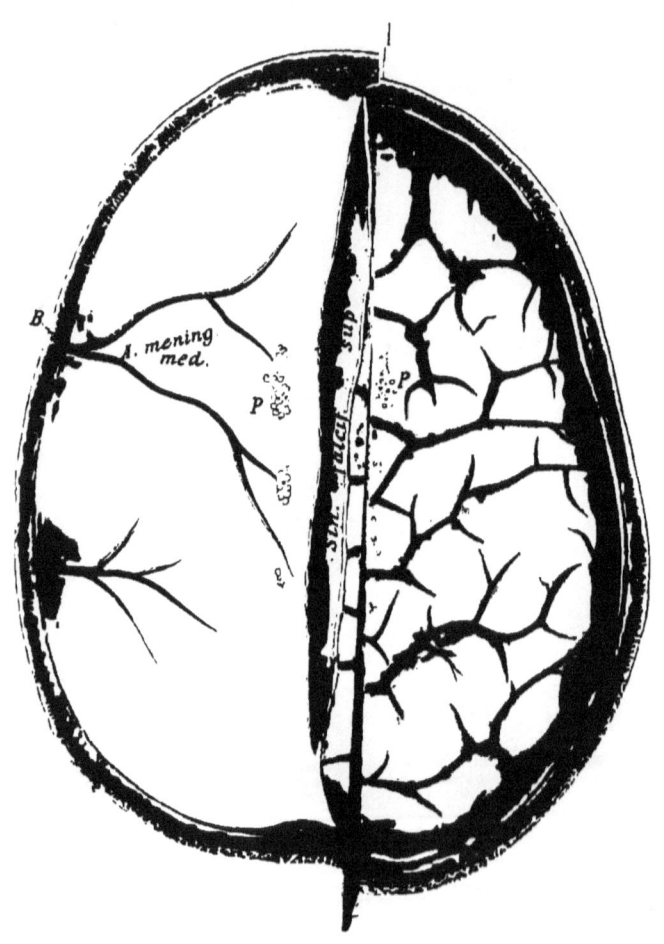

Explanation of Plate 1.

BRAIN *in Situ* SEEN FROM ABOVE AFTER REMOVAL OF THE SKULL CAP.

The brain is covered on the left with the dura and with the branches of the middle meningeal artery coursing in the bone channels. In the middle, the superior longitudinal venous sinus is open. On the right, the dura being removed, are the meninges, the arachnoid and beneath it the pia. The veins of the pia (in blue) empty into the sinus.

If the meninges, pia and arachnoid, are removed (lift the top segment on the right), the convolutions and sulci of the cerebral hemisphere come into view.

If one removes by a horizontal cut the upper portion of the hemisphere (lift top segment on left), the white substance of the hemisphere which lies centrally and which on cross section is known as the centrum semiovale comes into view, surrounded by the cortical substance forming the convolutions. Between the left and the right hemispheres can be seen at this level the corpus callosum, forming the connection between the medullary substance of the two sides (chordæ transversales, striæ longitudinalis on the surface of the corpus callosum). If one cuts more deeply into the medullary substance laterally to the corpus callosum (lift second segment on the left) he plunges directly into a cavity filled with fluid, the lateral ventricle, which is called specially at this point the cella media of the ventricle. From the bottom of the lateral ventricle the basal ganglion of the cerebral hemisphere projects up into the ventricular cavity (the corpus striatum). Behind the corpus striatum the vascular choroid plexus passes laterally. Of the hemispheres there have been cut through the frontal lobes, the central convolutions, the parietal lobe, the occipital lobe; each of these lobes with more or less of its convolutions has been involved.

Explanation of Plate 2.

HORIZONTAL CROSS SECTION OF THE HEMISPHERES AT THE LEVEL OF THE CORPUS CALLOSUM (THE CORPUS CALLOSUM REMOVED).

The corpus callosum ($c.c.$) is cut through at its anterior end, which bends under to form the knee, and at its posterior thickened end, the splenium, and removed. Beneath it in the middle the fornix appears, consisting of two limbs. Anteriorly the space between the knees of the corpus callosum and the anterior section of the fornix which bends toward the base, is filled by two small laminæ, the septum pellucidum ($s.p.$), between which is an interval, the ventricle of the septum. The lateral ventricle is more exposed by a deeper removal of the substance of the hemispheres. One then recognizes its anterior horn (cornu anterior, $c.a.$) in the frontal lobe, its posterior horn ($c.p.$) in the occipital lobe, and its inferior horn in the (to be considered transparent) temporal lobe in the depths of the hemisphere. Between the diverging occipital lobes the tentorium, a prolongation of the dura covering the cerebellar hemispheres and the vermis, comes into view.

The posterior pillars of the fornix rise out of the inferior horn from the free borders of a pad lying there, the cornu Ammonis, as the tænia hippocampi or fimbria. If the pillars be cut off in front and the entire fornix turned back the anterior pillars of the fornix can be seen (crus descendens fornicis) passing into the deeper parts. Under the fornix there lies outstretched a prolongation of the pia, the blood-vessel-carrying tela choroidea media with the median choroid plexus in the middle arranged as a pair of columns. These lead, passing closely beneath the anterior pillars of the fornix (foramen of Munro), into the lateral ventricles, and from here further into the inferior horn as the lateral choroid plexus.

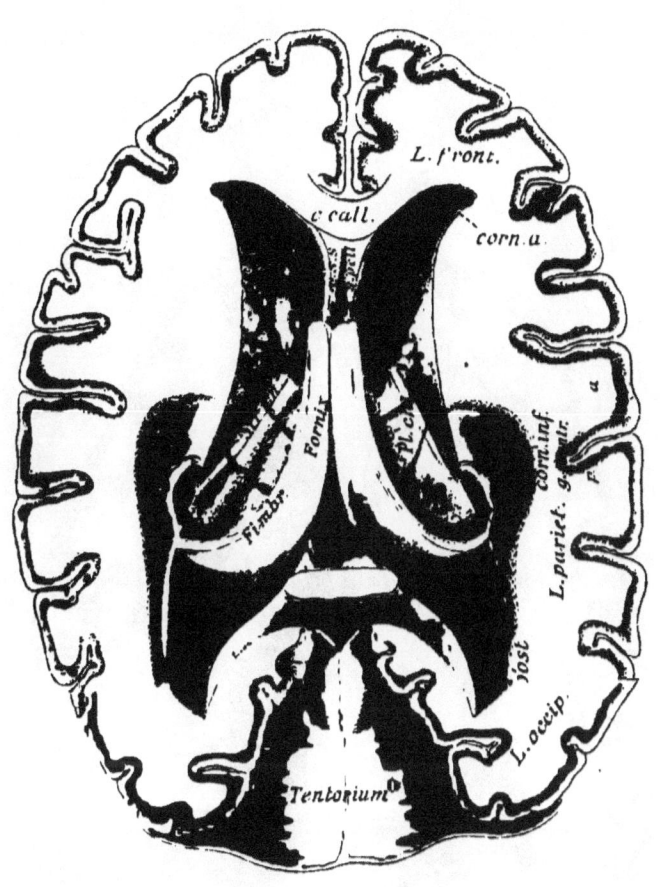

Tab. 3.

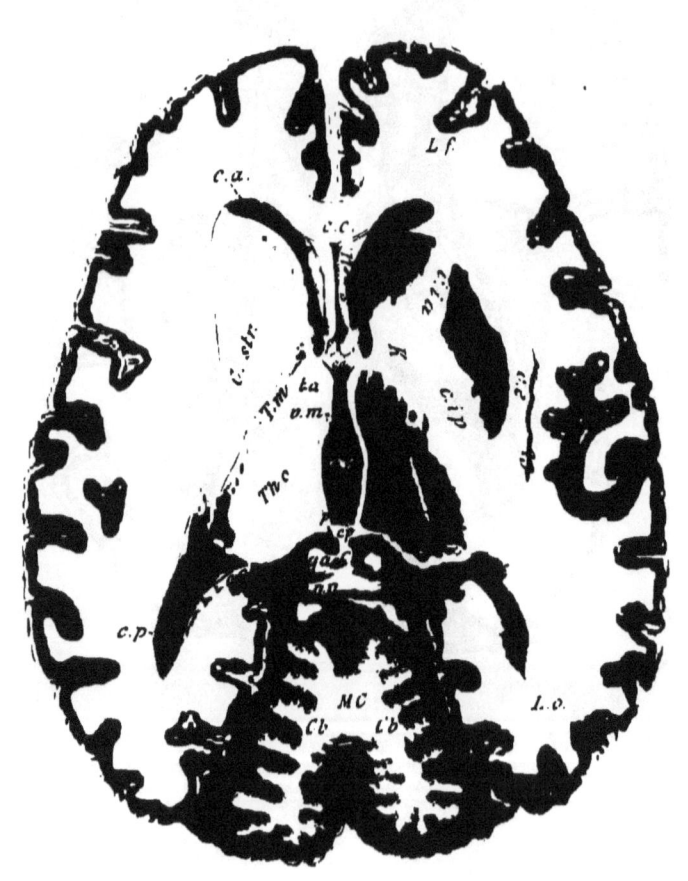

Explanation of Plate 3.

HORIZONTAL SECTION OF THE BRAIN AFTER THE THIRD VENTRICLE HAS BEEN LAID OPEN.

(Right side deeper than left.)

After removal of the choroid plexus and the posterior pillars of the fornix there can be seen in the median line the cavity of the third (middle) ventricle limited on both sides by the optic thalamus; on the left the surface of the corpus striatum lies bare (the head in the anterior, the tail in the posterior horn); between it and the thalamus the small stria cornea lies. The thalamus shows anteriorly its tuberculum anticum ($t.a.$) and posteriorly as a prominence, the pulvinar.

More deeply and in front of the descending pillars of the fornix lies the anterior commissure ($c.a.$) passing in a transverse direction; behind, in the middle of the third ventricle, is the median commissure ($c.m.$), and at the posterior end the posterior commissure. Over the latter the tænia thalami forming the medial thalamus unites with that of the other side. Here the pineal gland is attached (conarium, c). Behind the thalamus lie the corpora quadrigemina, anterior and posterior. Behind that the cerebellum is seen cut through superficially, the worm (vermis) in the middle, the hemispheres laterally, the white medullary substance centrally, the gray cortical substance peripherally.

On the right the head and the tail of the caudate nucleus are seen on cross section; the middle portion, which passes convexly, is carried away. There is also to be seen the thalamus on cross section after removal of the upper layer of the same. Latterly therefrom the lenticular nucleus appears, which together with the nucleus caudatus forms the corpus striatum. The inner section of white medullated substance which is thus divided off is called the internal capsule ($c.i.$), the lateral section, the external capsule ($c.e.$).

Explanation of Plate 4.
HORIZONTAL SECTION OF THE BRAIN THROUGH THE
CENTRAL GANGLIA.
(Left higher, right deeper.)

On the left (deeper than in Plate 3 on the right) the frontal lobe, the central gyri, the parietal lobe, the occipital lobe (on the right the temporal lobe) are cut through. At the level of the fossa of Sylvius lie the convolutions of the island of Reil (insula).

The medullary matter between the lenticular nucleus and the caudate nucleus is called the anterior limb of the internal capsule ($c.i.a$); that between the lenticular nucleus and the optic thalamus the posterior limb ($c.i.p$). Both limbs unite at the so-called knee of the capsule.

Laterally to the external capsule, close to the convolutions of the island of Reil, lies the claustrum (Cl).

The lenticular nucleus is divided into an outer portion, the putamen; and into several inner portions, forming the globus pallidus.

On the right the lenticular nucleus has become broader, the thalamus has disappeared, and in its place lies the subthalamic region with the red nucleus ($n.r.$) and the subthalamic body ($c.sth$). The limbs of the capsule have separated, the posterior having transformed into the peduncle of the cerebrum. Lateral therefrom the lateral (gl) and medial geniculate bodies (gm) are cut through.

In the medullary substance of the cerebellum, the corpus dentatum cerebelli ($c.d.$) can be seen, and on the left the nucleus tegmenti. On the right the fourth ventricle is opened beneath the vermis, the brachia conjunctiva (B) pass from the region of the corpora quadrigemina to the cerebellum.

After complete removal of the cerebellum (pull back the cover) the surface of the medulla comes entirely into view, forming the floor of the fourth ventricle (fossa rhomboidalis). It is bounded by the following cerebellar medullary tracts, which are here cut through: the processus cerebelli ad corpora quadrigemina (brachia conjunctiva, B), the processus cerebelli ad medullam oblongatam (restiform bodies, $r.r.$), and between both lies the processus cerebelli ad pontem passing toward the base.

Tab. 4.

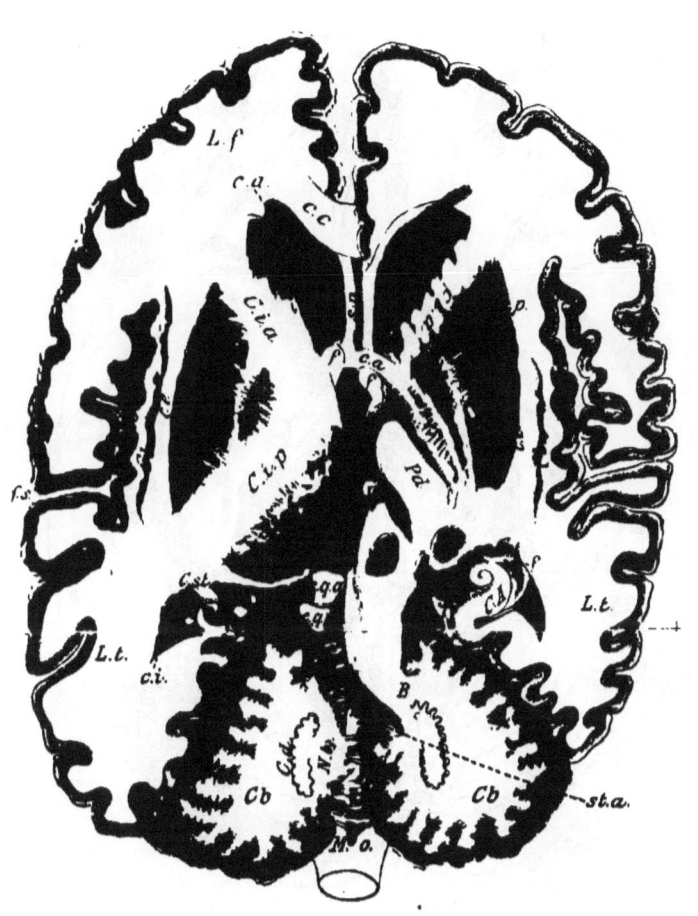

Tab. 5.

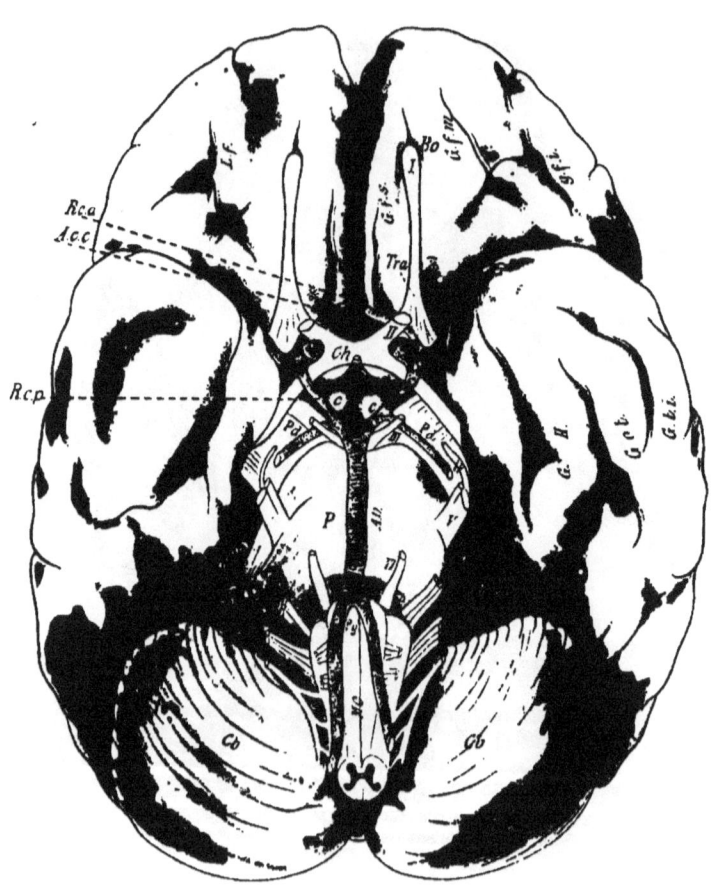

Explanation of Plate 5.

BASE OF THE BRAIN WITH THE NERVES AND BLOOD-VESSELS.

The deep fossa of Sylvius separates the frontal from the temporal lobes.

The crusta passes out of each hemisphere (pedunculus, *Ped.*) and enters with its fellow of the other side into the broad pons, covered by the transverse processus cerebelli ad pontem of the same. At the posterior end of the pons the pyramids (*Py*), the continuation of the peduncles, and near them the olives come into view, thus forming the base of the medulla. The occipital lobes of the cerebral hemispheres are covered by this portion (brain stem) and by the cerebellum (*Cb*). Between the cerebral peduncles lie the corpora candicantia (*c c*) (mamillary bodies), and the infundibulum, which merges into the hypophysis in the sella turcica. From the optic tracts (*tr.o.*) arises the chiasm of the optic nerves (*Ch*) and from the latter the optic nerves. Laterally therefrom are the trigonum, the olfactory tract, and the olfactory bulb (*I*). The position of the twelve cranial nerves is apparent without further consideration. They are: II. Optic nerve. III. Motor oculi. IV. Trochlear. V. Trigeminal. VI. Abducens. VII. Facial. VIII. Auditory. IX. Glosso-pharyngeal. X. Pneumogastric. XI. Spinal accessory. XII. Hypoglossal. All of these up to the fourth arise from the base. The arteries at the base form the circle of Willis; they are the vertebral (*A.v.*), the basilar (*A.b.*), the deep cerebral (*R.p.*), the internal carotid (*C.i*), the anterior and posterior communicating branches (*R.c.a.* and *p.*), the artery of the fossa of Sylvius (*A.f.S.*), and the artery of the corpus callosum (*A.c.c.*).

The *crusta* is covered after its entrance into each hemisphere by the marginal convolutions of the temporal lobes (gyrus hippocampi), the anterior end of which is known as the uncus. Laterally therefrom passes the occipito-temporal gyrus (indicated as *G. H.*). The convolution designated as *G.o.t.* is the inferior temporal gyrus.

Explanation of Plate 6.

FIG. 1.—*Base of the Brain after Removal of the Brain Axis and Cerebellum.*

The brain stem has been severed by a section through the cerebral peduncles made at right angles to the corpora quadrigemina, which lie dorsally therefrom.

The section thus obtained is divided into two parts, a basal segment (crusta, peduncle) and a dorsal segment (tegmentum), and between the two the substantia nigra ($S.n.$). The optic tracts starting from the chiasm pass around the peduncles further forward beneath the (to be considered transparent) temporal lobe and end laterally to the corpora quadrigemina in the lateral geniculate body ($c.g.l.$).

The olfactory tract (removed on right side) passes into the trigonum olfactivum.

FIG. 2.—*A Survey of the Projection Tracts.*

The medullary masses of the centrum semiovale have their origin (or end) in the brain cortex. A portion of this passes in the internal capsule further downward. These end partly in the optic thalamus (brown), but the greater part goes further and passes either through the crusta or the tegmentum.

The most important portion of the peduncular tracts are the motorial pyramidal tracts (red); they course further through the brain stem and spinal cord. The most important part of the tegmental tract is the fillet tract (green). This also passes further as far as the end of the medulla. Medullary tracts come likewise from the cerebellum, which go to the spinal cord (restiform bodies) and to the brain stem (tegmentum). All of these tracts connect by their prolongations the cortex of the brain and cerebellum with the periphery of the body, and taken together they are known as the projection tracts of the cortex.

Tab. 6.

Fig 1

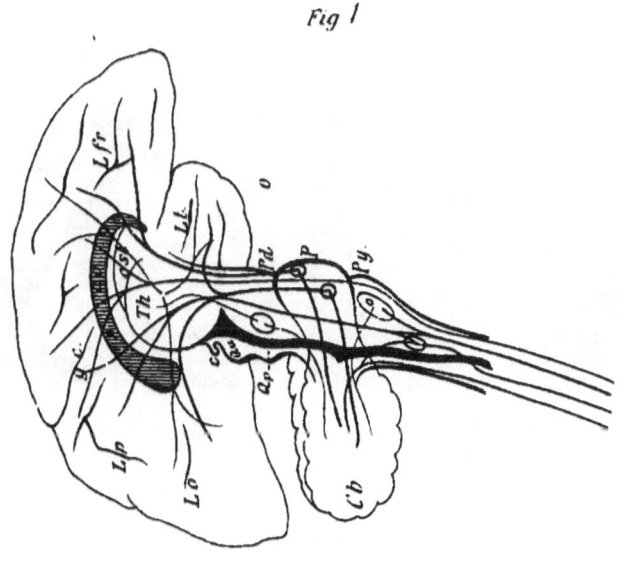

Fig 2

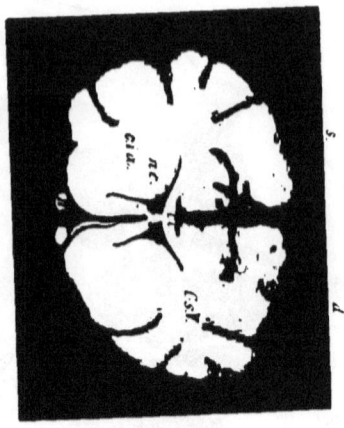

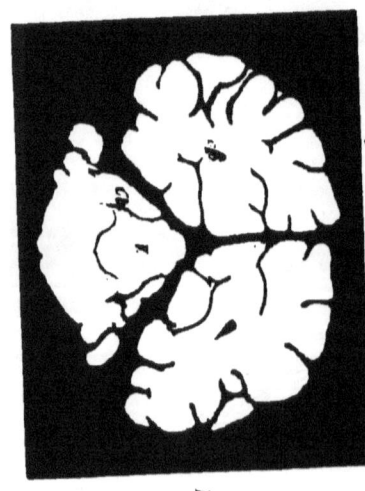

Explanation of Plate 7.
FRONTAL SECTIONS THROUGH THE BRAIN OF A DOG.
(Photographs.)

One recognizes without further explanation the central white medullary substance and the convoluted gray cortex of the cerebral hemispheres. The medullary substance is, comparatively, much smaller than in man and the convolutions less numerous.

FIG. 1.—*Section through the Frontal Lobes.*

It includes the head of the nucleus caudatus which projects into the lateral ventricles. The corpus callosum and the septum pellucidum beneath it are easily recognized, as well as the section of the anterior limb of the internal capsule ($c.i.$). The optic nerve is seen on the base and laterally to it the olfactory lobe (atrophied in man).

FIG. 2.—*Section through the Optic Thalamus (Th), and the Tail of the Caudate Nucleus.*

Under the corpus callosum ($c.c.$) the fornix (fimbria). Lateral from the thalamus the internal capsule (posterior limb), which passes downward to the peduncle (optic tract, corpora candicantia, inferior horn of lateral ventricle in temporal lobe).

FIG. 3.—*Section through the Cerebral Peduncle.*

The crusta (*Ped.*) is liberated from its hemispherical envelopment, above is the tegmental area separated from the former by the substantia nigra. Laterally from this the optic tract ends in the lateral geniculate body ($g.l.$) and the pulvinar. The third ventricle extends under the posterior commissure as the aqueduct of Sylvius.

FIG. 4.—*Section through the Occipital Lobes, the Cerebellum, and the Medulla.*

In the occipital lobes the posterior horn of the lateral ventricle; under the same, in the middle, the worm of the cerebellum; laterally therefrom parts of the hemispheres. In the medulla cross-section, brachium conjunctivum, floor of the fourth ventricle, fillet area, pyramids, and pons fibres.

Explanation of Plate 8.

FRONTAL SECTIONS THROUGH THE HUMAN BRAIN STEM.

(Photographs.)

FIG. 1.—*Section through the Third Ventricle.*

The corpus callosum, fornix, and lateral ventricle are recognized at once, likewise the cross section of the caudate nucleus and beneath it the optic thalamus (anterior tubercle). Laterally therefrom the internal capsule (posterior limb). Further outward and below the internal portion of the lenticular nucleus (globus pallidus) and a smaller portion of the outer segment (putamen). The optic tract, the mammillary bodies, the lowest convolutions of the temporal lobes (uncus of the hippocampal gyrus), nucleus amygdale (*n.a.*), and the inferior horn of the lateral ventricle.

FIG. 2.—*Section Behind 1 through the Mammillary Bodies.*

Corpus callosum, fornix, thalamus, caudate nucleus, internal capsule (which has in part passed out as peduncle), lateral ventricle, third ventricle, hippocampal convolution, optic tract.

To the corpora candicantia there passes a bundle from the thalami. The descending pillars of the fornix end laterally in the corpora candicantia.

FIG. 3.—*Section through the Posterior End of the Third Ventricle*, shortly before its transition to the aqueduct; the subthalamic region lying beneath the optic thalami comes in contact with the other side. Here are the red nucleus, the fillet (lemniscus), and the subthalamic body.

The crusta is free, and laterally to it is the lateral geniculate body with the beginning of the optic tract.

Tab. 8.

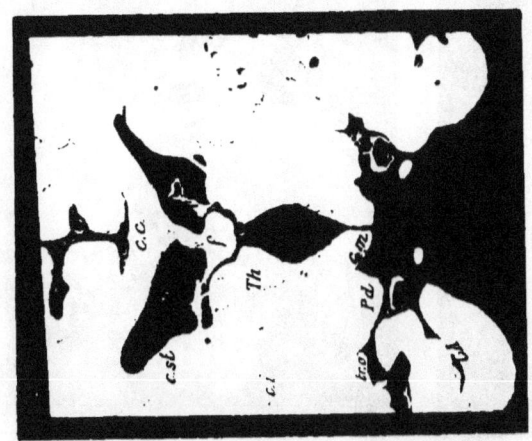

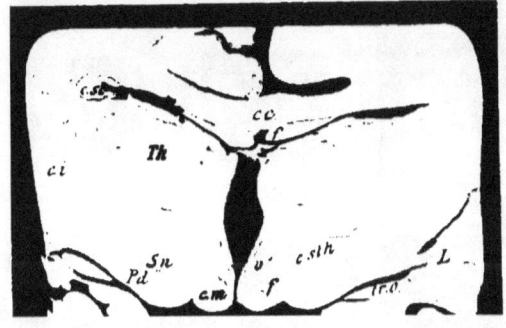

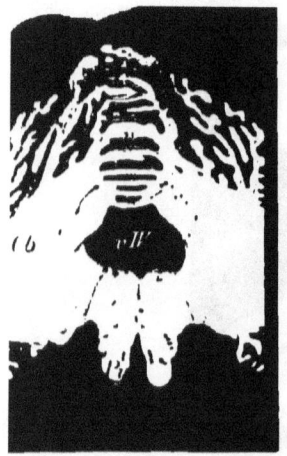 

Fig. 3.   Fig. 4.

Fig. 6.   Fig. 7.

Fig. 5.   Fig. 8.   Fig. 9.   Fig. 10.

Explanation of Plate 9.

FRONTAL SECTION THROUGH THE BRAIN STEM; PARALLEL TRANSVERSE SECTIONS THROUGH THE MEDULLA AND SPINAL CORD.

(Photographs.)

FIG. 1.—*Section through the Anterior Corpora Quadrigemina.*

Under the corpora quadrigemina (*qa*) the aqueduct of Sylvius (*A*) and laterally to it the medial geniculate body (*gm*); under it the tegmentum of the cerebral peduncles (red nucleus [*nr*]), fillet (*L*), and beneath (substantia nigra) the crusta. Beneath the aqueduct lies the motor oculi nucleus.

FIG. 2.—*Section through the Middle of the Pons.*

The larger portion of the fibres of the peduncles end partly in the pons, while a part course farther as pyramidal tracts, covered by the transverse fibres of the pons. The brachium conjunctivum passes out of the tegmentum as the lateral wall of the fourth ventricle (which results from enlargement of the aqueduct of Sylvius) to the cerebellum. The roof of the fourth ventricle is formed by the vermis of the cerebellum, and laterally therefrom are the convolutions of the cerebellar hemispheres.

FIG. 3.—*Section through the Fourth Ventricle at the Posterior End of the Pons.*

For the boundaries of the ventricle see 2. The brachium conjunctivum has entered the medullary substance of the cerebellum. The hemispheres of the cerebellum (*Cb*) are cut through at their greatest diameter (medullary and cortical substance). The pyramids (*Py*) pass out free; caudad therefrom the principal tegmental region (fillet, facial nucleus, auditory).

FIG. 4.—*Section Immediately Behind Fig. 3.*

The processus cerebelli ad medullam oblongatam (restiform bodies), which in the previous section united cerebellum and medulla, have here passed entirely out of the medullary substance; within the medullary substance is seen the corpus dentatum cerebelli (*cd*); close by the pyramids (*Py*) are the olives. Uvula (*u*) of the worm (*v*), nodules (*n*) (lobes of the cerebellum).

FIG. 5.—*Section through the Posterior End of the Floor of the Fourth Ventricle (Calamus Scriptorius).*

The restiform bodies approach each other. The cerebellum is partly cut away.

FIG. 6.—*Section through the Nuclei of the Posterior Columns, Passing through the Restiform Bodies (fp).* Pyramids (*Py*), proton of lateral columns (*fl*).

FIG. 7.—*Section at the Level of the Pyramidal Decussation.*

Each pyramid passes in great part into the lateral column of the opposite side.

FIGS. 8, 9, 10.—*Transverse Section of Cervical Cord, Dorsal Cord, and Lumbar Cord.*

H-shaped central gray substance (anterior horn, posterior horn). The anterior, lateral, posterior columns (*fa, fl, fp*) make up the white substance.

Explanation of Plate 10.

TRANSVERSE SECTION OF THE SPINAL CORD *in Situ.*

(Photographs.)

The usually white medullary substance is here colored black (medullary sheath stain). See text, Section VI.

FIG. 1.—*A Child's Cervical Cord with its Roots in the Spinal Canal* (Sixth Cervical Vertebra and its Environs).

In the bony capsule within the three membranes lies the cord. The anterior roots (*ra*) (motorial) spring from the anterior half of the central gray substance (anterior horn); the posterior roots (*rp*) (sensory) pass out from the posterior horns. Both roots, after uniting, pass out of the spinal canal through the intervertebral foramen (*fi*) as a peripheral nerve (*Np*). The posterior root forms in the foramen the intervertebral ganglion (spinal) (*gi*). Body of vertebra (*v*), spinous process (*Pr sp*), spinal arch (*av*).

FIG. 2.—*Lumbar Cord of a Newborn within its Membranes.*

The outer envelope is the dura (*dm*) surrounded by Breschet's plexus of veins (*Plv*), the middle the arachnoid (*ar*); both of these surround the cord loosely. The inner envelope, the pia (*pm*), with the blood-vessels surrounds the cord closely. In the subarachnoidal and subdural space is the cerebro-spinal fluid. The anterior (*ra*) and posterior (*rp*) roots pass off from the spinal cord in the lumbar region at a lower level than that at which they arise. They then unite and surround the cord (on deeper transverse sections called the cauda equina). In the spinal cord the deep anterior longitudinal sulcus (*sa*), the shallow posterior longitudinal sulcus, (the anterior columns (*fa*), the lateral columns (*fl*), the posterior columns (*fp*), in the middle the central canal, (*cc*) (the continuation of the ventricular system), anterior and posterior commissures between the anterior horn (*Ca*) and posterior horn (*Cp*) of the gray (central) substance.

Tafel 10.

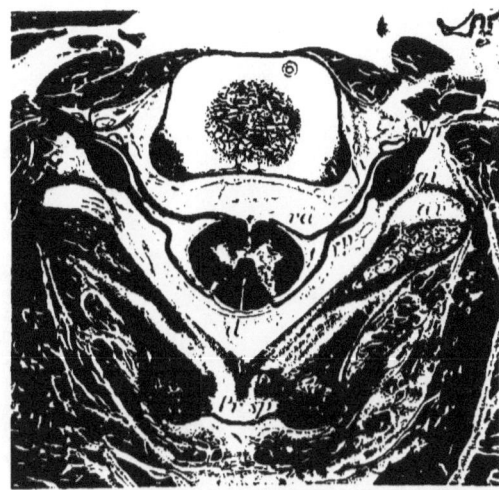

Fig. 1.

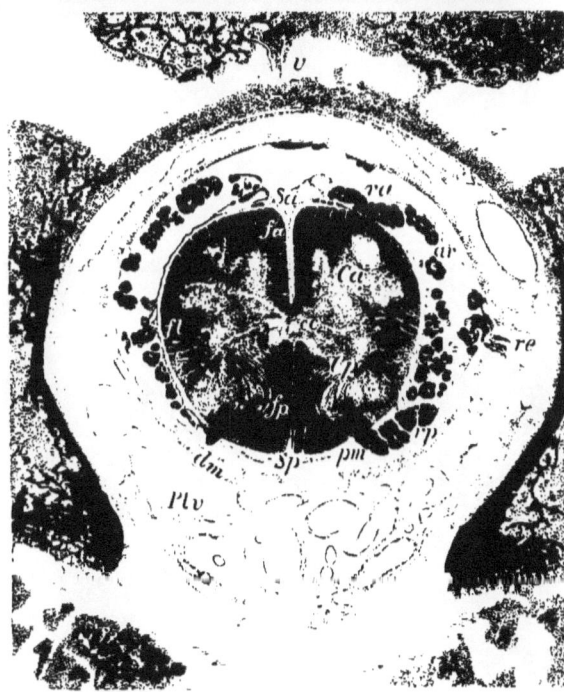

Fig. 2.

II.

DEVELOPMENT AND ARCHITECTURE OF THE NERVOUS SYSTEM.

(Section II. of Textual Part.)

Tab. 11.

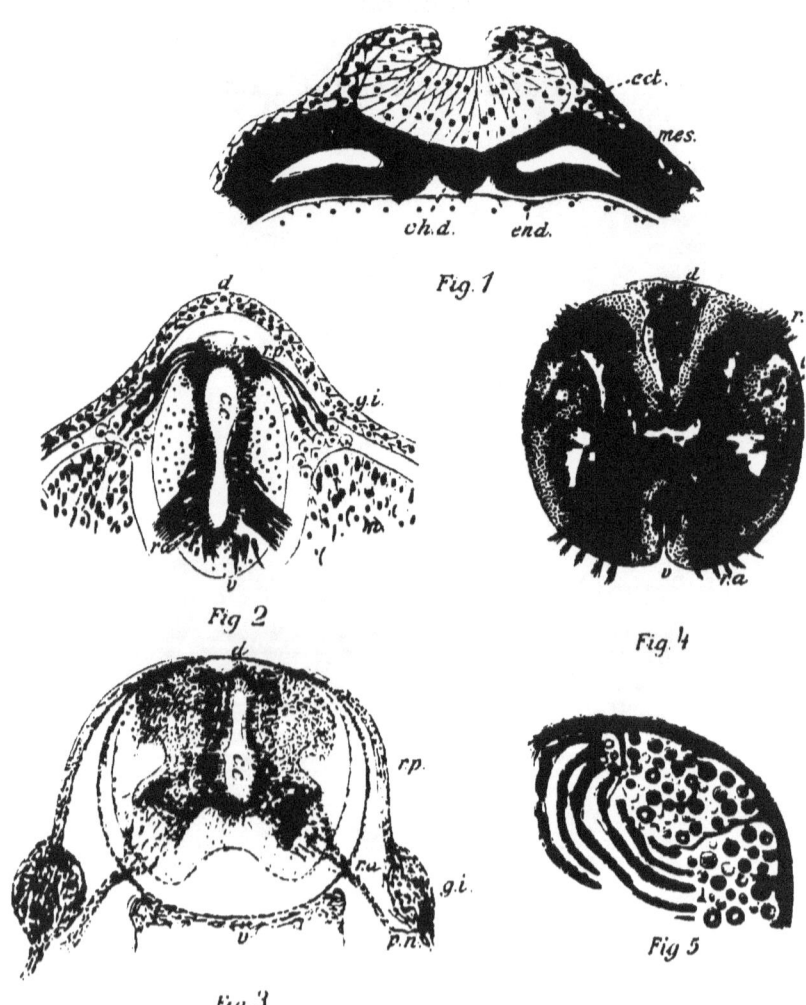

Explanation of Plate 11.

EMBRYONAL PRIMARY CONSTITUENT OF THE ORGANS.

FIG. 1.—*Cross Section through the Medullary Groove.*
The medullary groove (yellow) originates from the ectoderm; laterally from it is the ganglionic ledge, from the cells of which the intervertebral ganglia develop.

Mesoderm (blue), chorda dorsalis (brown), entoderm (red), ectoderm.

FIG. 2.—*Cross Section through the Medullary Channel (First Month).*
The stratified epithelial cells of the walls differentiate themselves into spongioblasts (connective-tissue framework substance) and neuroblasts (nervous substance proper). From the neuroblasts develop ventrally (anterior horn proton) the anterior motor roots (ra), dorsally from the cells of the spinal ganglia (gi) the sensory posterior roots grown into the dorsal part of the cord.

FIG. 3.—*Cross Section of the Spinal Cord (Second Month).*
Differentiation of the central and peripheral gray substance and later of the white substance. The anterior and posterior horns have formed. The roots as in Fig. 2.

FIG. 4.—*Cross Section of Spinal Cord (Eighth Month).*
Medullation of nerve fibres in the white substance follows successively and systematically. Tracts which are anatomically or functionally in relationship become medullated simultaneously.

The pyramidal tracts ($Py.l.$ and $a.$) are the last to get their medullary sheaths. In the figure the fibres that are clothed with medullary sheaths are stained black, the non-medullated are unstained, brownish.

FIG. 5.—*Peripheral Nerve of a Newborn.*
Here likewise the medullation of individual nerve fibres occurs successively, but in its entirety earlier than in the central organs. The medullary proton is here not complete and definite.

Explanation of Plate 12.

EMBRYONAL DEVELOPMENT OF THE BRAIN.

FIG. 1.—*Brain and Spinal Cord at the Third Month.*

The fore-brain (*I*) has already separated into the two hemispheres and covers the tween and middle brain (*III*) (optic thalamus and corpora quadrigemina proton). Behind is the hind-brain (cerebellar proton) (*IV*) and the after-brain (medulla proton) (*V*). In the spinal cord the two enlargements (cervical and lumbar swellings) (*J.C.* and *J.L.*).

FIG. 2.—*Base of the Brain* (*Fourth Month*).

The frontal (*L.f.*) and the temporal lobes (*L.t.*) are separated by the Sylvian groove. Chiasm (*II*), cerebral peduncle (*Pe*), pons (*P*) are recognizable. Cerebellum (*Cb*), pyramids (*Py*), olives (*o*).

FIG. 3.—*Median Surface of Brain* (*Fourth Month*).

Corpus callosum (*c.c.*) and fornix are clearly recognizable; anteriorly between the two the septum pellucidum (*S.p.*). The descending pillar of the fornix bends posteriorly and ends in the mammillary body.

FIG. 4.—*Outer Surface of a Cerebral Hemisphere* (*Fifth Month*).

On the surface begins the fissure formation. From above downward is the central sulcus of Rolando. From below the fossa of Sylvius (*f.S.*), in the depths of which is the proton (*J*) of the island of Reil.

For further reference to medullary sheath proton in the brain see Plate 48.

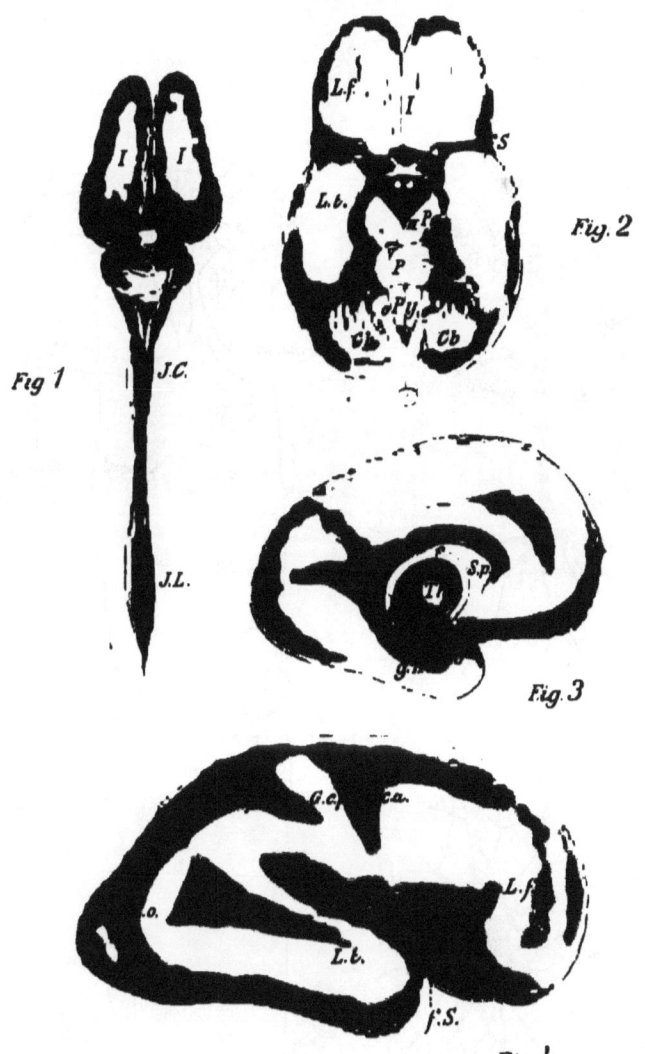

Tab. 13.

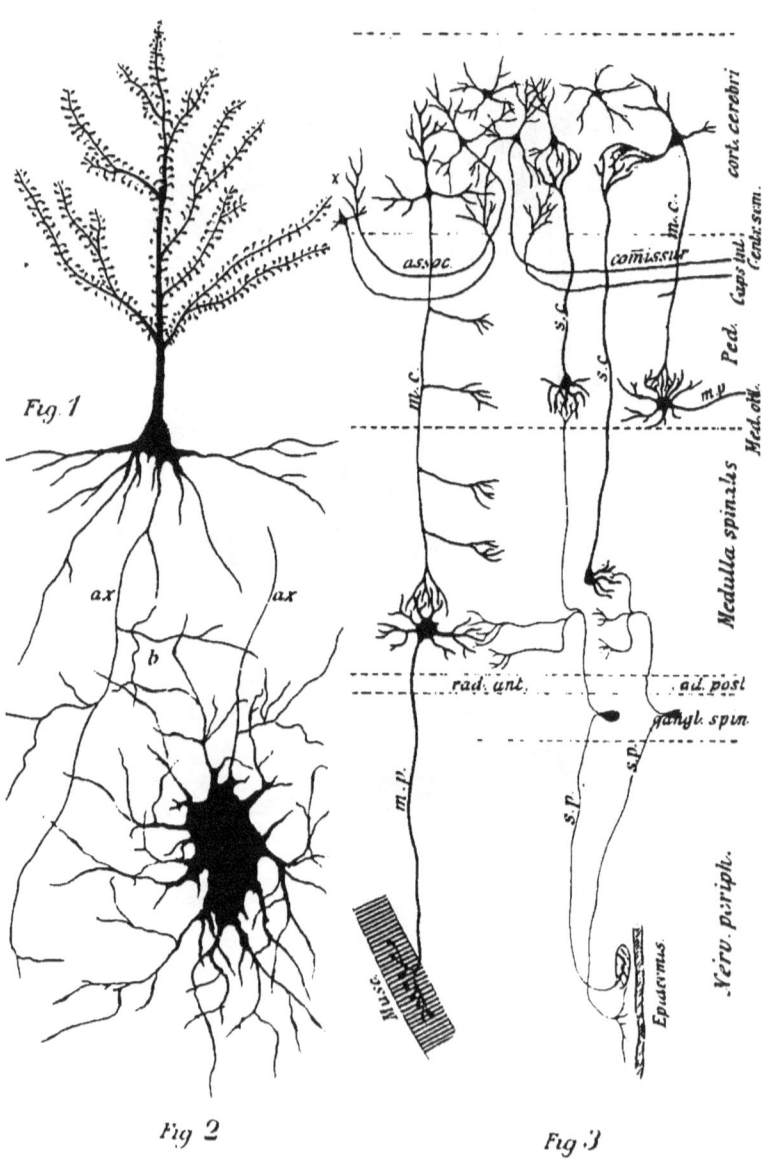

Fig. 1

Fig. 2

Fig. 3

Explanation of Plate 13.
STRUCTURE OF THE NEURONS.

Every neuron consists of a nerve cell and its prolongations, the nerve fibres. The cells as well as the fibres ramify multifariously. Two kinds of neurons are differentiated: neurons with a long nerve prolongation (Deiters' type) and neurons with only a short nerve prolongation (Golgi's type).

FIG. 1.—*A Pyramidal Cell from the Cerebral Cortex with its Tree-like Ramifications* (dendrites, protoplasmic prolongations). Each prolongation has numerous bud-like, later twig-like branches. The never fibre of the cell (axis-cylinder prolongation) is designated by a. Its terminal ramification occurs remotely in another part of the brain or of the spinal cord and takes place around the dendrites of ganglion cells there lying (Fig. 2). This neuron has its course and termination entirely in the central organ, and is therefore called central neuron.

FIG. 2 represents a ganglion cell with its dendrites from the anterior horn of the spinal cord. Between its ramifications lies the terminal arborization of the nerve fibre of the central neurons. Its nerve fibre (a, axis-cylinder prolongation) after a short course goes out of the central organ and passes peripherally to the soft parts. The entire neuron is therefore called peripheral neuron.

FIG. 3 shows the schematic construction of the nerve tracts from such neurons, especially the motorial (cortico-muscular tract) is represented above. The central motor neuron (red) splits up around the cells of the peripheral neuron (blue); similarly in part is the construction of the sensory tract from two neurons. The peripheral sensory neuron (brown) ends with its terminal arborization around the cells of the central sensory neurons (green). The fibre arborization in the centrifugally conducting motorial neurons occurs in a reverse sense to that of centripetally conducting sensory. These are then the projection tracts. In the cortex are other neurons (green), which from their manifold connections are called the association tracts. (More in text, Section III.).

Explanation of Plate 14.

Ganglion Cells.

Nerve cells (ganglion cells) are always found collected in more or less large groups in certain places called gray substance. They are most numerous in the cortex, the central ganglia (corpus striatum, *i.e.*, caudate and lenticular nucleus and the optic thalamus), corpora quadrigemina, corpora geniculata, in the nuclear layer of the tegmentum, and in the gray substance around the ventricles from the aqueduct of Sylvius on, which substance continued into the spinal cord is known as the anterior and posterior horns.

Fig. 1. Only the coarser prolongations of the cells are recognized by this method of staining; in spite of that the cell structure is in these clearer, many having pigment.

(*a*) Pyramidal cells of the cerebral cortex.

(*b*) Cells from the anterior horn of the spinal cord.

(*c*) Cells from Goll's and Burdach's nuclei (nuclei of the posterior columns).

(*d*) Cells from the substantia nigra, strong pigmentation.

(*e*) Cells from the corpus striatum, optic thalamus, claustrum.

Fig. 2. A cell of Purkinje from the cerebellar cortex. These cells have the richest ramifications of all the ganglion cells (photograph).

The significance of the dendrites is not yet completely understood (whether of purely nervous or nutritive function).

Tab. 15.

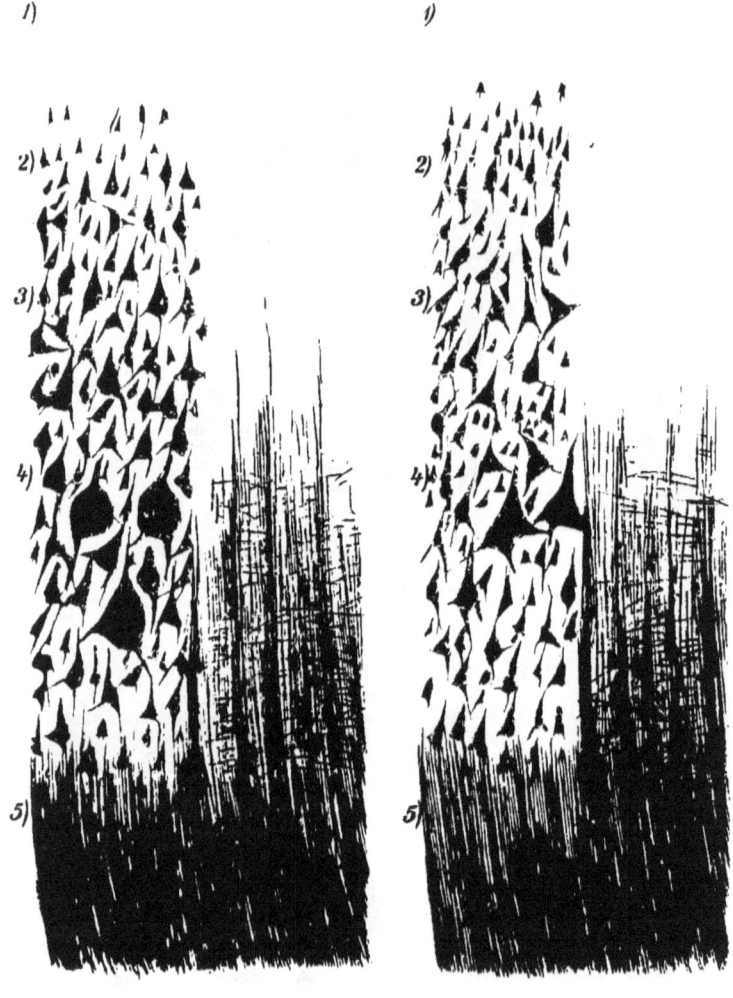

Fig. 1   Fig. 2

Explanation of Plate 15.

CEREBRAL CORTEX.

FIG. 1 *shows the Cortex of the Frontal Lobe.*
FIG. 2 *that of the Occipital Lobe, Strong Magnification.*
The left half shows the pyramidal cells, the right half the fibres; both naturally lie together.

The cortex of different convolutions is individually very differently constructed, nevertheless there is certain uniformity in the proton.

In the layer of pyramidal cells there can be differentiated an upper, a middle, and an inferior formation. The upper layer has small pyramidal cells; the middle, large, contiguous, pyramidal cells. The largest cells are found in the middle layer (especially numerous in the frontal lobes and the central convolutions). In the inferior layer smaller cells are found.

The nerve fibres of the cortex come in part from the medullary substance of the hemispheres (*5*). These radiate, arranged in tufts, in a straight manner into the cortex, and end somewhat beyond the middle of the cortex (radii). Between these vertically passing medullary fibres ramify the remaining cortical fibres, thus forming the tangential fibre network, a part (*2, 3*) above the radiations, superradiating, a part within the radiation, interradiating network. Between both (*3*) is a thick collection of tangential fibres (Baillarger's, Gennari's, Vic d'Azyr's layer), which can be seen distinctly in the cortex of the cuneus. Close beneath the pia and very superficially lies (*1*) a layer of such tangential fibres. In it and beneath it until the first pyramidal cell layer is reached and also in that are seen small oval cells (not represented) with short ramifications (Golgi's type). All these cells are built on the same principle as the pyramidal cells of Plate 13, Fig. 1. The entire tangential network is in connection with the ramification of the fibres. The medullary radiations are in connection in part in the cortex with the nerve fibres that arise from its cells, and in part with fibres that come from one portion of the cortex and undergo terminal distribution in another, such as the projection and association fibres. See scheme of the cortex, Plate 13, 3, above.

Explanation of Plate 16.

BRAIN CORTEX AND MENINGES.

FIG. 1.—*Scheme of the Meninges.*
The dura (*d*) covers over the bony tables of the skull cap, representing the position of the periosteum; beneath is the subdural lymph space.

The arachnoid (*a*) united with the pia (*p*) lies in close apposition to the brain and passes over the fissures like a bridge. Beneath it the multi-compartment subarachnoidal space in communication with the ventricular cavity, which forms large spaces (filled with cerebro-spinal fluid) at the base in front of and behind the pons. The Pacchionian bodies (*G.P.*) are connective-tissue accumulations of the arachnoid. The pia internally lies on the outer layer of the cortex and sends its blood-vessels in all over the cortical substance; it covers, therefore, the sulci as well. Beneath the brain cortex is the radiating medullary substance.

FIG. 2 is a tuft of the choroid plexus, the vascular prolongation of pia in the third ventricle. One recognizes the cubical epithelium and the vessel walls.

FIG. 3 is a single small cortical convolution from the cerebrum; the radiating black-colored (naturally white substance) medullary substance passes out bush-like (photograph).

FIG. 4 shows the somewhat different architecture of the cortex of the cerebellum compared with the cortex of cerebrum. The medullary substance (black) is smaller. Close to it is the nuclear layer (zona granulosa) of the cortex (blue), on the borders of which are the large cells of Purkinje (see Plate 14, Fig. 2). Their ramifications form, as well as those of other medullary fibres, the molecular zone of the surface. Then comes the closely apposed pia covering (not represented). A portion of the nuclear cells form (not represented) short ramifications of Golgi's type.

Tab. 16.

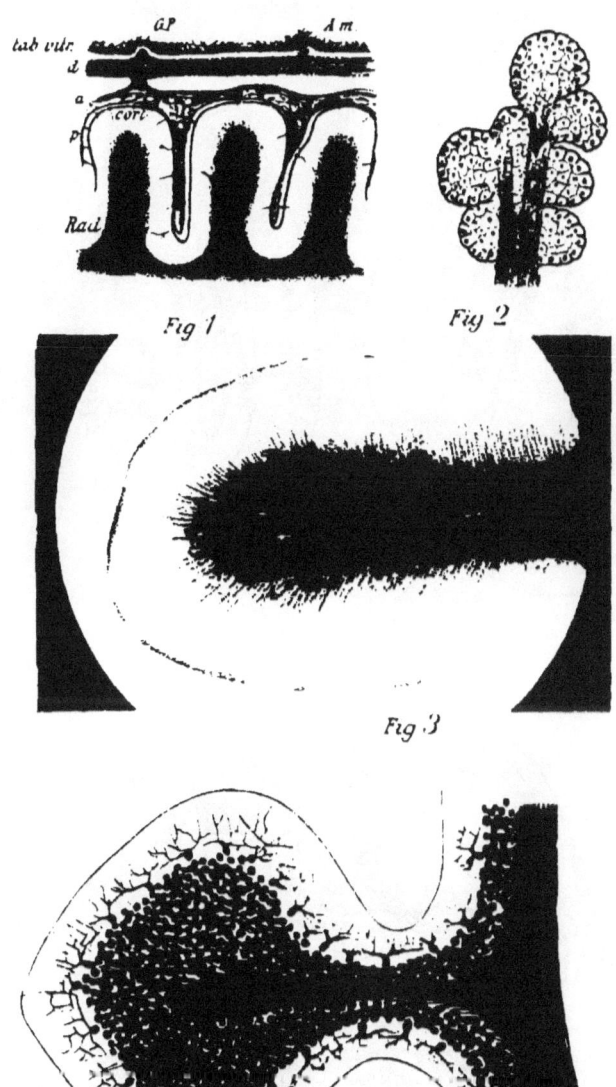

Tab. 17.

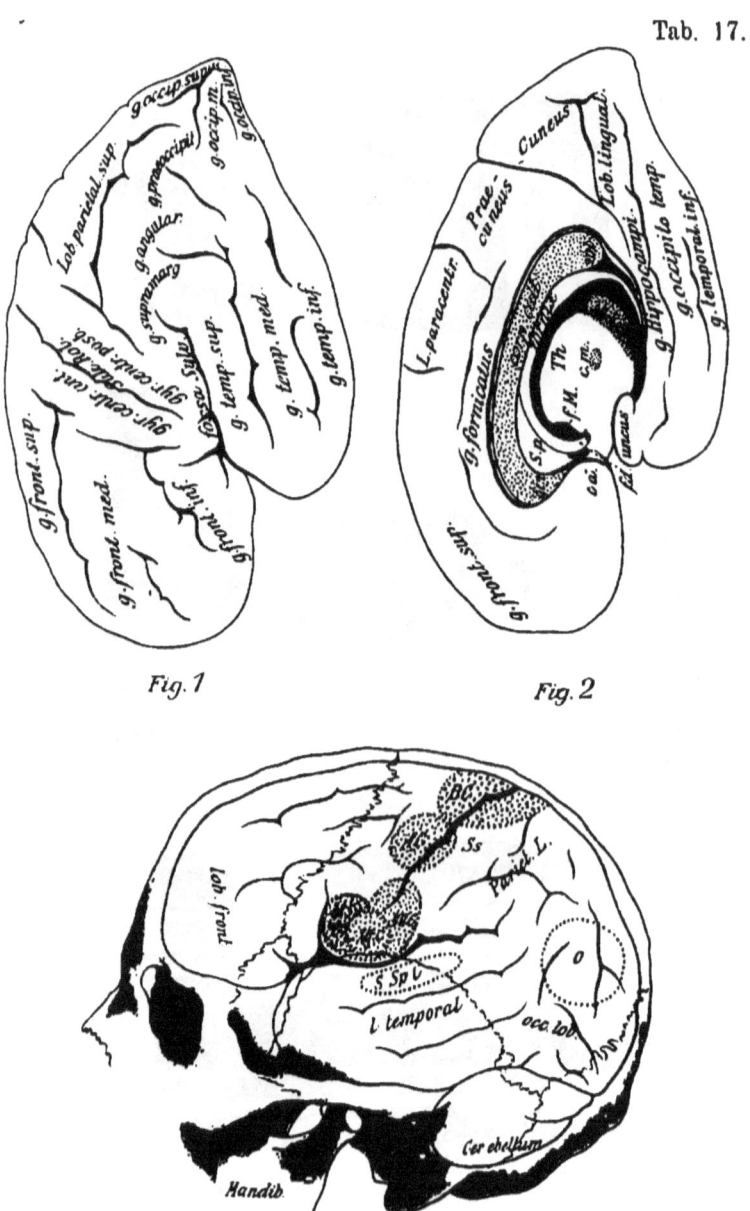

Explanation of Plate 17.
CEREBRAL CONVOLUTIONS.

FIGS. 1 and 2 show schematically the convolutions and fissures of the cortex on the external and median surfaces of the hemispheres.

One recognizes in the frontal lobe three gyri (gyrus superior, middle, and inferior). These are separated behind from the central convolutions (*g. centr. ant.* and *post.*) by the central sulcus. On the median surface the lobus centralis is called the lobus paracentralis. The parietal lobe divides into the upper and lower group of convolutions, the lower being divided into the supramarginal gyrus, the angular gyrus, and the pre-occipital, and on the median surface the upper appears as the precuneus. In the occipital lobes are the three gyri— superior, middle, and inferior—and on the median surface there is found the cuneus and lingula separated by the calcarine fissure. The temporal lobes have likewise three convolutions, the superior, middle, and inferior gyri, and on the median surface the temporo-occipital gyrus borders on the inferior marginal convolution, the gyrus hippocampus (anteriorly the uncus). The upper marginal convolution constitutes the gyrus fornicatus.

FIG. 3.—*The Position of the Psycho-Motorial and Psycho-Sensorial Cortical Centres in the Skull Cavity.*

The central representation for motility and sensibility of the periphery of the body is in the cortex of the cerebral hemispheres. Parts known with certainty to be the seat of such representation are the central convolutions, the inferior frontal and the upper temporal lobes, and the cortex of the occipital lobe (especially its median surface). The apportionment and limitations are given in the figure. BC motor centre for the lower extremites, AC for the upper, $VIIC$ for the face musculature, $XIIC$ for the hypoglossal; $MSpC$ motorial, $SSpC$, sensorial speech centre, SC centre for vision; Ss, sensory sphere (trigeminus and extremities). For further, see Section IV. The upper end of the central sulcus (fissure of Rolando) is to be found 5 cm. behind the coronary suture. A sagittal section through the root of the nose and the external protuberance passes somewhat beneath the fossa of Sylvius. With this line the central sulcus forms an angle of about 65°. If the so formed Rolando's line is divided into three parts, the position of the motor centres will be practically determined.

Explanation of Plate 18.

FIG. 1 shows the position of the motor and sensory nuclei from which the cranial nerves arise or in which they end. They are delineated in the brain stem and medulla as viewed from above. From the blue nuclei arise the motor, from the green the sensory cranial nerves. The cells of the green nuclei form the beginning of the central sensory neurons (compare Plate 13). The Roman letters indicate the cranial nerves: *Vn* nasal descending (probably motor), *Vc* caudal descending root of the trigeminus; *Vm*, motor, *Vs*, sensory trigeminus nucleus; *VIIIc.*, cochlear nerve of auditory, *VIIIv.*, vestibular nerve of the auditory.

FIG. 2 shows various isolated nerve fibres which collectively constitute the white medullary masses of the brain, the columns of the spinal cord, and the peripheral nerves, but are found also in the gray substance. The most important part of each nerve fibre is the axis cylinder (cell prolongation). It is surrounded by a medullary envelope of varying thickness, the white medullary sheath which is put on in segments around this. Within the peripheral nerves around this is the thin sheath of Schwann.

(*a*) Longitudinally running medullated nerve fibres, the medullated substance stained black and covering the axis cylinder. This is seen better on transverse section (*b*) and in the other fibres (*c*) and (*d*). The axis cylinder is made up of numerous individual fibrillæ.

Tab. 18.

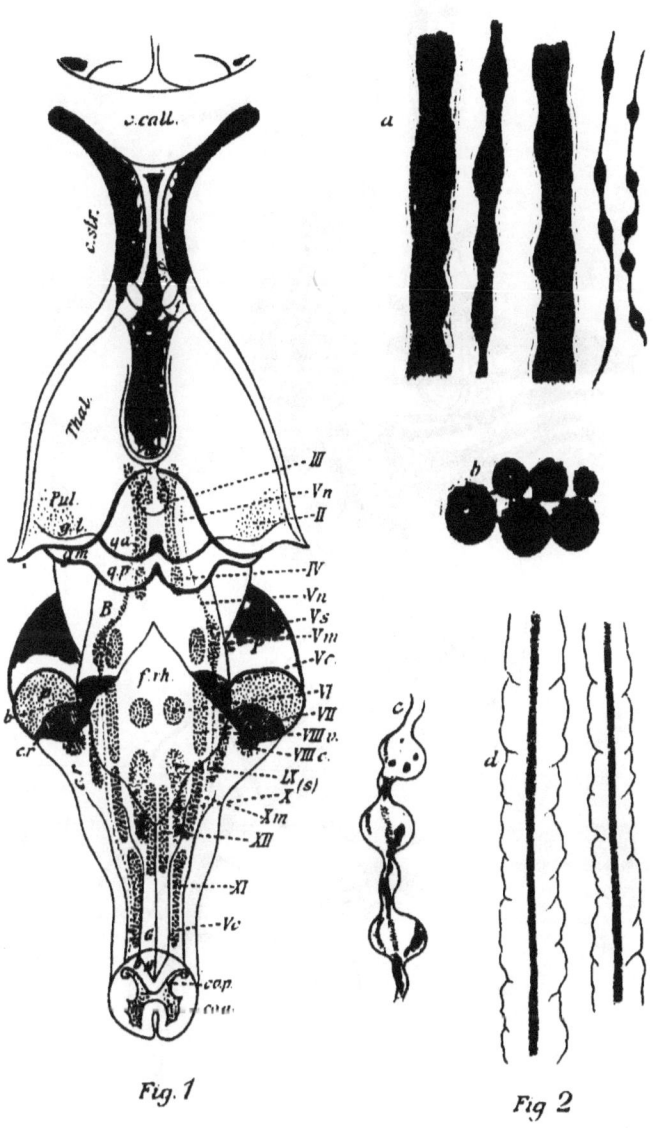

Fig. 1 Fig. 2

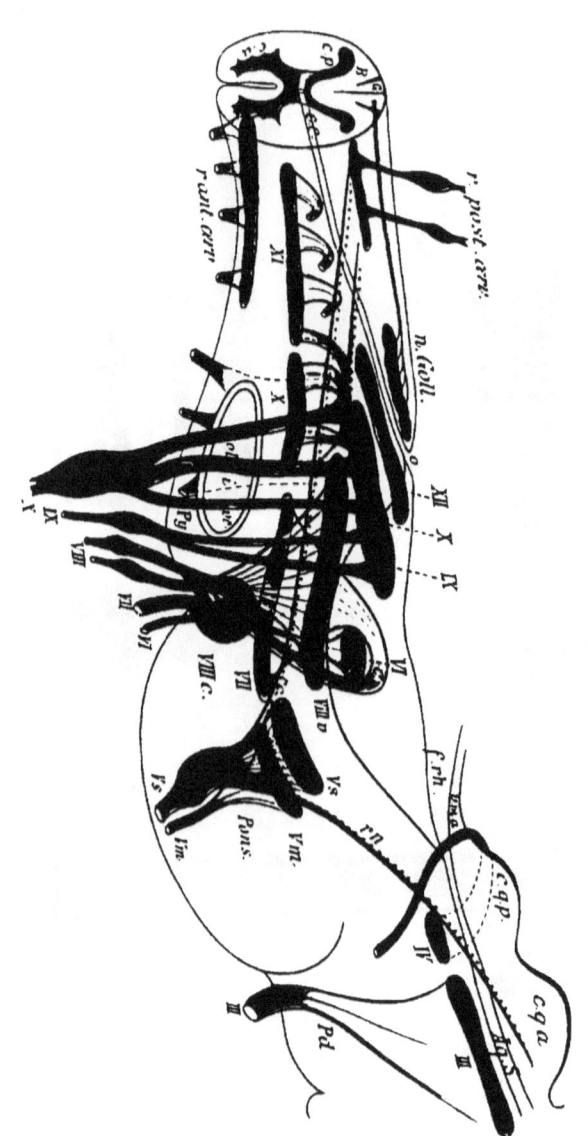

Explanation of Plate 19.

LATERAL VIEW OF THE MEDULLA OBLONGATA WITH THE SCHEMATIC REPRESENTATION OF THE NUCLEI AND THE INTRAMEDULLARY COURSE OF THE CRANIAL NERVES.

The uppermost nerve is the motor oculi (III). Its nucleus, of long extent and consisting of several parts, lies in the central gray matter of the ventricle beneath the aqueduct of Sylvius. Behind it lies the nucleus of the trochlear nerve; the nerve (IV) passes dorsally as one of the cranial nerves out behind the posterior quadrigeminal bodies, decussates with its fellow of the other side, and then passes to the base.

The trigeminus (V) arises in two portions; the anterior motorial (Vm) arises from the motor (V) nucleus, to it is joined the probably motorial nasal descending root (rn), which arises laterally from the aqueduct (here represented as sensory root). The posterior sensory (Vs) originates from the Gasserian ganglion and separates in the pons in three portions: one portion ends in the sensory vagus nucleus; a second portion passes as the caudal ascending root (rc) through the medulla as far as the first cervical segment, where it ends by degrees; a third portion passes to the cerebellum. The facial (VII) arises from its extensive and deeply situated nucleus in the medulla, gathers itself together to the knee of the facial, passes around the adducens nucleus (VI) with an outward curvature, and finally emerges beneath it. The course of the abducens is simple. The auditory (VIII) ends as the cochlear nerve ($VIIIc.$) in the ventral nucleus, and as vestibular nerve ($VIIv$) in the dorsal nucleus and its environs; one portion passes as descending root of the VIII, a short distance further toward the spinal cord (not represented). The sensory glossopharyngeal-vagus fibres come out from the petrosal and jugular ganglia and end in the sensory IX and X nuclei in the posterior end of the calamus scriptorius (X, green); a portion passes as descending root of the X some distance toward the spinal cord (solitary fascicle). The motor fibres arise in part from the nucleus ambiguus (X, blue), and in part from the spinal accessory nucleus.

The hypoglossal (XII) nucleus lies in the caudal dorsal segment of the medulla and sends its fibres out through the medulla laterally to the pyramids. The nucleus of the spinal accessory (XI) reaches far down in the upper cervical cord and its fibres pass out through the lateral columns; $r.p$=posterior, upper cerviacl root from the spinal ganglia, $r.a$=anterior upper cervical roots.

Explanation of Plate 20.

THE NUCLEAR LAYER OF THE PERIPHERAL MOTOR NEURONS.

FIG. 1 shows a segment from the anterior gray horn of the cervical cord with its motor ganglion cells. From these cells the roots of the anterior motor nerves originate (*ra*), which pass through the adjoining white substance of the antero-lateral columns (the medullated nerve fibres stained black on transverse section).

The medullated fibres which are woven around the cells subserve in part the transportation of reflexes (reflex collaterals from the posterior roots).

FIG. 2 represents a part of the hypoglossal nucleus. The numerous large multipolar cells give rise to the fibres of the hypoglossal nerve. Coursing around the cells are innumerable medullated nerve fibres of varying diameter which form a fine network. These fibres are in part central tracts, and in part unite the nuclei of the two sides, especially the inner parts (in this way the unison of action of both halves of the tongue is explained) and thus serve for transmission of the reflexes. In an entirely similar way the facial, the motor oculi, etc., nuclei are constructed.

In these two territories, therefore, begins the peripheral motor neuron, while simultaneously the terminal ramifications (the latter are not perceptible with this staining) of the central neuron take place about these cells.

Tab. 20.

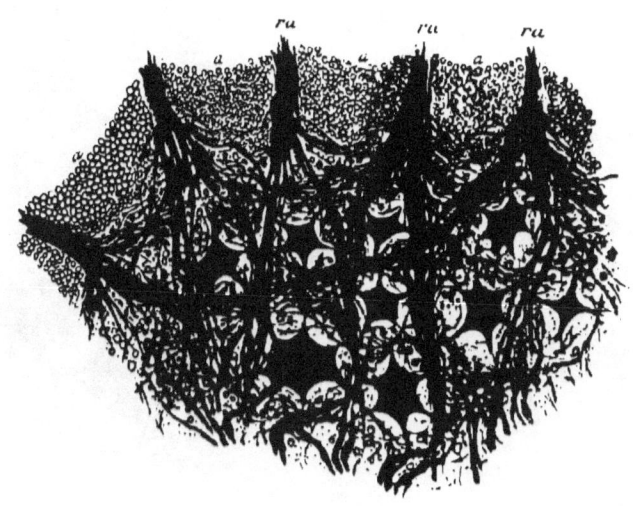

Fig 1

Fig 2

Tab. 21.

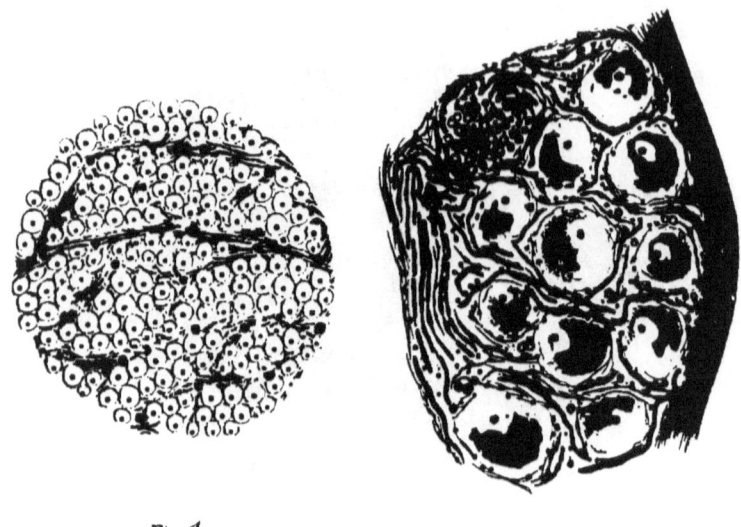

Fig. 1

Fig. 2

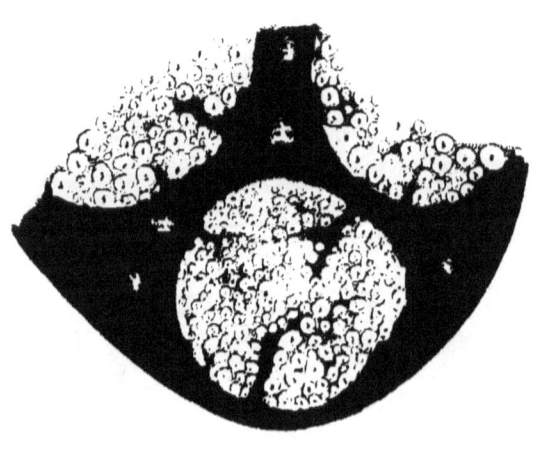

Fig. 3

Explanation of Plate 21.

Fig. 1.—*Transverse Section through the White Medullary Substance of the Spinal Cord (Lateral Column).*

The medullary substance consists of the closely woven medullated nerve fibres which are embedded in the framework of glia. As the preponderating number of nerve fibres course longitudinally it follows that on cross section we see mostly the fibres cut across (sun images). In the centre of each fibre is the axis cylinder.

Fig. 2.—*Section through a Spinal Ganglion.*

The strongly pigmented cells of the ganglion (likewise, for instance, the Gasserian ganglion) are surrounded by outgoing and incoming sensory nerve fibres, the outgoing fibres forming the posterior root. A fibre passes into each cell and out again (T formation). These cells are the places of origin of the peripheral sensory neurons.

Fig. 3.—*Section through a Peripheral Nerve.*

Every nerve is made up of several bundles. The entire nerve is surrounded by a connective-tissue envelope, the epineurium. The bundles lie in the several compartments of the perineurium. The epineurium separates the individual nerve bundles into sections. In these lie first the medullated nerve fibre surrounded by the sheath of Schwann in fibrillæ sheaths. Blood-vessels and lymph vessels ramify in the perineurium. In reality the nerves in transverse section are much closer together than they are represented here. One should remark the different thickness of the fibres.

Explanation of Plate 22.

Fig. 1.—*Schematic Representation of the Composition of the Cervical Cord on the Left, of the Lumbar Cord on the Right.*

Between the anterior fissure (sulcus anterior, *S.a*) and the anterior roots (*r.a.*) lies the anterior column. It consists of the anterior uncrossed pyramidal column (pyramidal tract, *Pya*) and the anterior ground bundle (*f.a.*). Between the anterior and posterior roots (*rp*) lies the lateral column. It is divided into the column of Gowers (*G*), the lateral pyramidal tract (*Cb*), the crossed pyramidal tract of the lateral column (*Py*), the lateral ground bundle (*fal*) (ground bundle of the lateral columns), and the lateral limiting layer (*fl*).

The posterior columns are divided into the columns of Goll (*G*) and the columns of Burdach (*B*). In the postero-lateral portion of the latter the posterior roots enter (the posterior root zone). At the apex of the posterior horn Lissauer's column is situated (*L*) (medullary bridge); *z. r.* = postero-external, *r.* = the ventral field of the posterior columns. The gray substance is divided into the anterior (*C.a.*) and the posterior (*C.p.*) horns; between them in the middle line lie Clarke's columns (*Cl*).

In front of the central canal (*c.c.*) lies the anterior white commissure (*C.a.*), and behind the same the posterior gray commissure. The ganglionic cells of the anterior horns are divided into a lateral and a medial group, each of which is further divided into an anterior and posterior portion. In the base of the posterior horn is the substantia gelatinosa (*S.g.*); contiguous with this is the zona spongiosa, and with the latter the above-mentioned white medullary bridge.

Fig. 2.—*The Central Canal and its Environments; Strong Magnification.*

The canal is lined with cylindrical epithelium. One recognizes the medullated fibres of the anterior and posterior commissures.

Tab. 22.

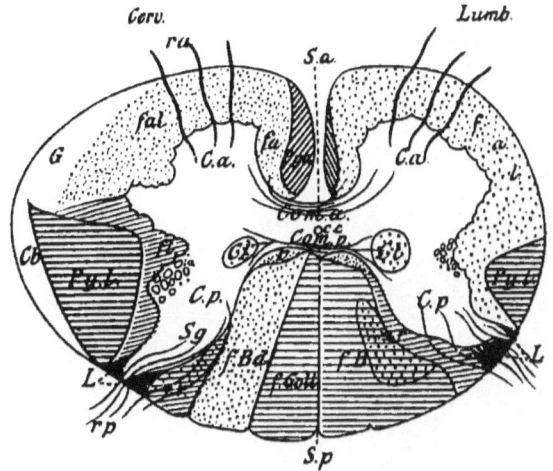

Fig. 1

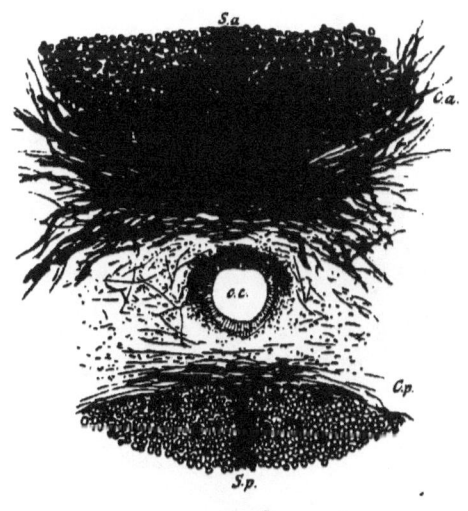

Fig. 2

III.
TOPOGRAPHICAL ANATOMY OF THE NERVOUS SYSTEM.

(Plates 23–48.)

Plates 49–52 schematic representation of the course of the fibres.

(SECTION III. OF THE TEXT.)

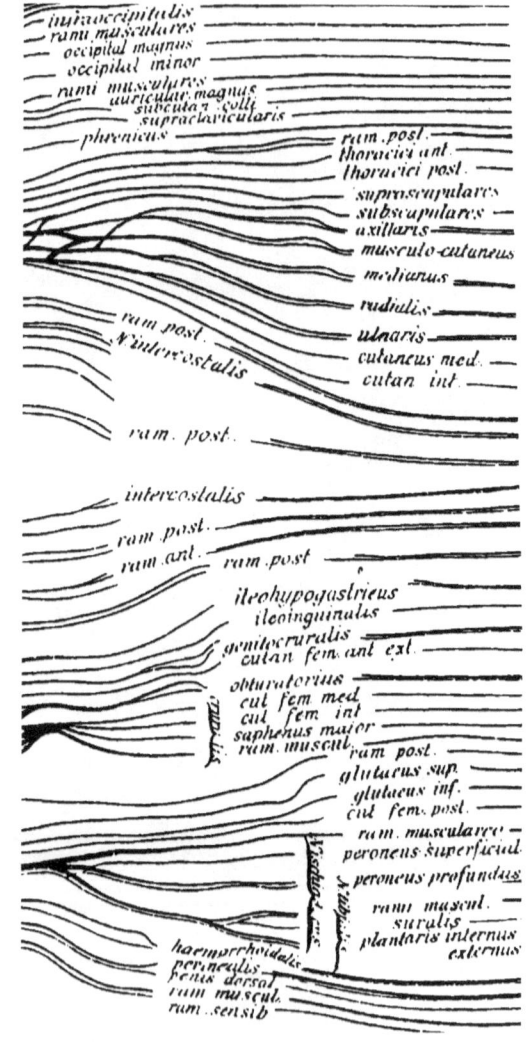

Serial Sections through the Central Nervous System.

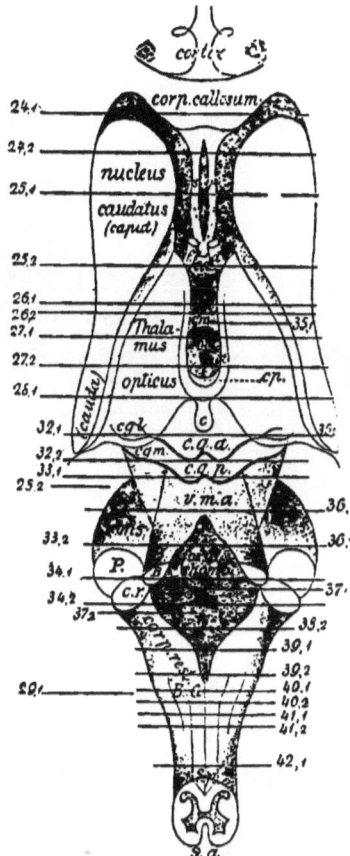

Plates 24 to 45 are from frontal sections, plates 29–31 from horizontal sections: first through the entire left hemisphere, then through the brain stem, the medulla, and the spinal cord.

The sections are prepared with medullary sheath stains; in part photographed, in part drawn, and are in no way schematic.

As the sections follow one another from the frontal region they represent in their continuity the continuation of the brain stem.

In all the sections the medullary substance is stained black, while the brain cortex is red, the subcortical ganglia and nuclear layers are stained blue. The amount of blue stippling will give one an idea as to the richness of cells.

Where it is not otherwise stated the drawings are of natural size.

One is able to reconstruct the relationship by comparisons with the previous and following section and to orientate himself by referring to the adjoining plan of division.

Remark. —The brain sections are from a boy sixteen years old. It must be remembered that in the fully developed adult the area of medullary substance is more developed in size and in detail.

Explanation of Plate 23.

DIVISION OF THE BRAIN AND SPINAL CORD NERVES.

One is able to recognize the level of a brain or spinal cord segment by the individual nerve roots which come from it. An exact representation of the segments of the brain stem and of the spinal cord are given in the text (Fig. 6, page 48). Of greatest importance is the relation of an individual segment of the spinal cord to the spinal column (spinous prolongations) and a knowledge of the point of exit of the spinal nerves. The nerves of the lumbar cord pass far below the segments from which they have their origin before they make their exit. This knowledge and a knowledge of plexus formation is presupposed.

The segments are represented in red; the motor nerves are blue, and the sensory after their exit from the plexuses are brown. In the same way the color of the lettering is represented. The further divisions of the sensory cutaneous nerves are given pictorially and more lucidly than by written description, in Figs. 10, 11, 12, text p. 102, and the following.

The motorial nerves (the most important) are numbered, and there is also a table setting forth the muscles that are innervated by these nerves and such essentials of their function as is carried by their names.

Tab. 24.

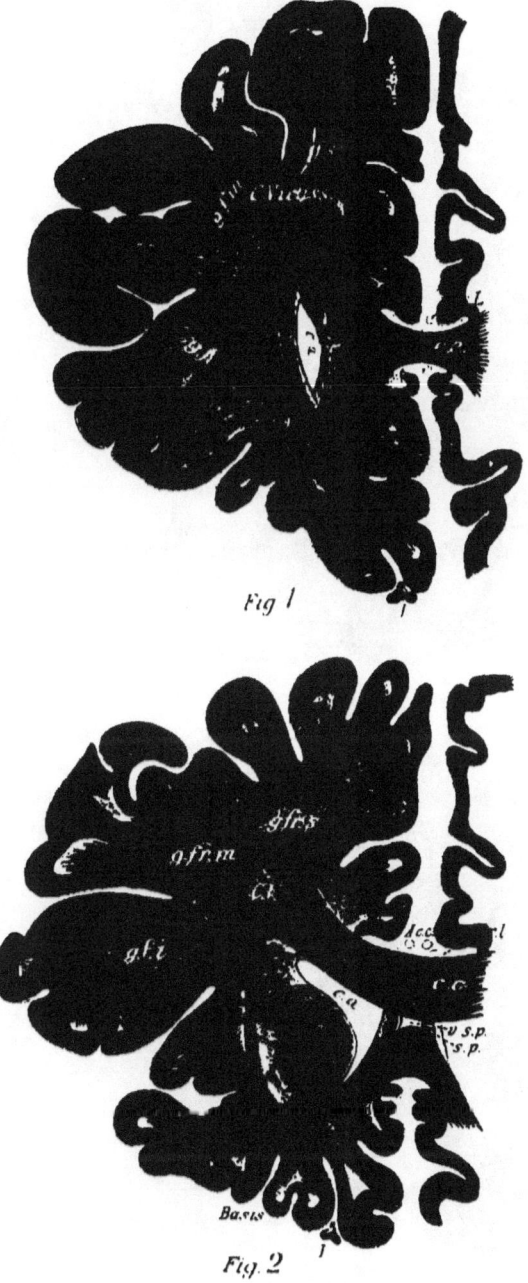

Fig 1.

Fig. 2.

Tab. 25.

Fig. 1

Fig. 2

Explanation of Plate 25.

FIG. 1.—*Frontal Section through the Middle of the Septum Pellucidum.*

In addition to the frontal lobes the apex of the temporal lobes (*l.t.*) is cut through. In the medullary substance the ganglion of the brain stem (corpus striatum) on account of the forward projection of the putamen (*L*) (the outer segment of the lenticular nucleus) is entirely involved. The anterior limb of the internal capsule breaks through the corpus striatum and thus divides it into the nucleus caudatus (*n.c.*) lying medially, and the nucleus lenticularis (*L*) lying laterally. Externally from the putamen is the external capsule (*c.e.*), and externally to this may be seen the claustrum (*Cl*). Septum pellucidum (*S.p.*), olfactory bulb (*I*) lateral ventricle (*v. l.*), corpus callosum (*c.c.*).

FIG. 2.—*Section through the Anterior Commissure.*

The ascending pillar of the fornix can be seen (*f*) passing up. The inner portion of the lenticular nucleus (*gl.p*), the globus pallidus, is likewise apparent. On the base the chiasm of the optic nerve (*Ch*) can be seen cut through laterally to the perforate substance, then the trigonum olfactorium (*I* instead of *II*). In the depth of the Sylvian fossa *f.s.* the anterior convolutions of the island of Reil (*Ins*) are included laterally to the claustrum. In the anterior portion of the temporal lobes begins the nucleus amygdalæ (*N. am.*); see following sections. The cortical convolutions have been cut away. Anterior commissure (*c. ant.*).

The third ventricle begins between the pillars of the fornix.

Explanation of Plate 26.

FIG. 1.—*Frontal Section through the Anterior Commissure.*

The descending pillars of the fornix (*f.d.*) pass backward to the mammillary body at the base. The body of the fornix continues beneath the corpus callosum, see Fig. 2. Between the adjacent cut-through portions of the thalamus (anterior nucleus) the third ventricle spreads out (*v.III*).

The globus pallidus (*gl.p.*) is seen to be made up of several segments and to surround with its basal fibres the internal capsule. From the chiasm the optic tracts (*tr.o*) pass backward to the lateral geniculate bodies. The anterior horn of the lateral ventricle (*c.a*) arches backward and is involved in the section which has been cut obliquely.

FIG. 2.—*Section through the Knee of the Internal Capsule.*

Beneath the posterior portion of the superior frontal convolutions are the central convolutions (*g.c.a.* and *p.*), and in the depths of the fossa of Sylvius the convolutions of the island of Reil are included. In the temporal convolutions the nucleus amygdalæ (*N.a.*) is included in the sections. Medially is the cornu ammonis (*c.am.*), the inner vaulting of the hippocampal gyrus. Between the caudate nucleus and the optic thalamus are the stria cornea. Around and through the internal capsule (*C.ip.*) the ansa lenticularis (*L.s.*) winds, passing from the globus pallidus. A portion of its fibres unite and pass further dorsally beneath the territory of the thalamus to the fillet area (upper fillet).

In the thalamus (*Th*) the lower stalk of the thalamus passes from the temporal lobes beneath the ansa lenticularis. The thalamus has become very much larger and projects upward (tuberculum ant.) as far as the pillar of the fornix, and in it one recognizes different fibre pathways (laminæ medullares). Tuber cinereum (*Inf.*)

Tab. 26.

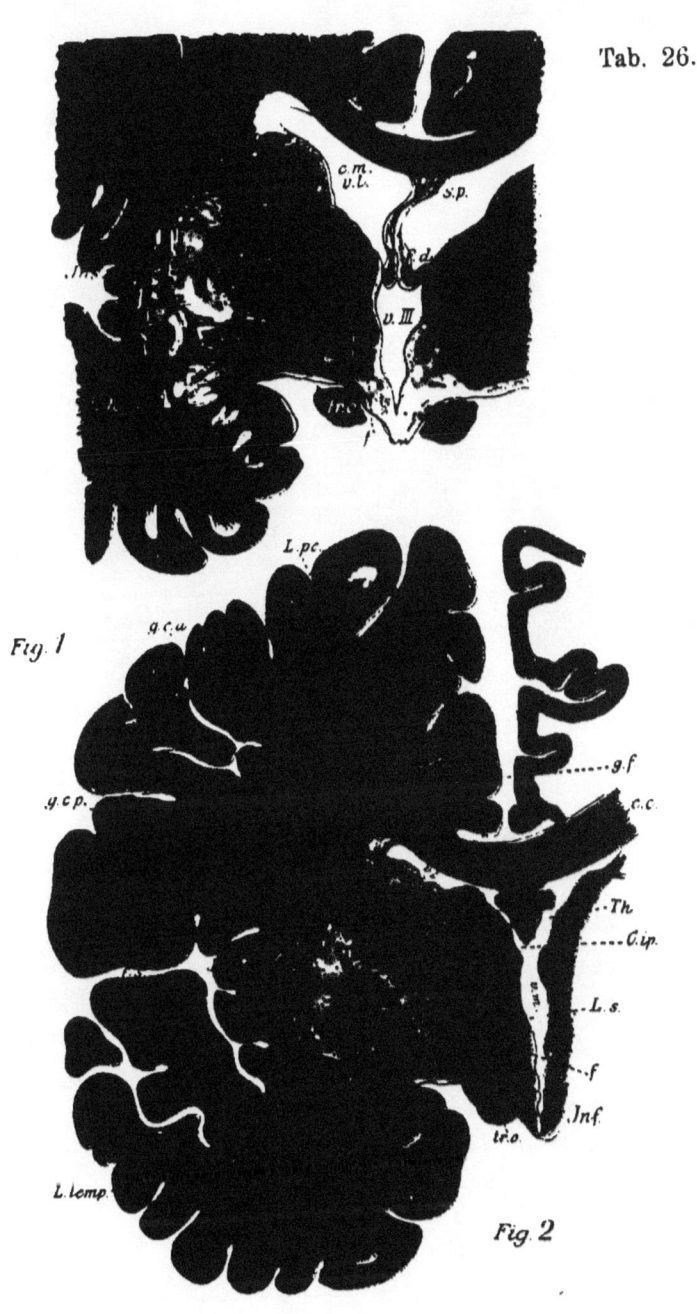

Explanation of Plate 28.

FIG. 1.—*Section through the Parietal Lobe* (anterior portion).

The brain stem passes farther and farther out from the hemisphere. The peduncle (*Ped.*) is entirely free; the thalamus (pulvinar) is connected only by its upper and posterior stalk. Between the peduncle and the internal capsule (posterior segment) the lateral geniculate body (*c.g.l.*) (in which the optic tract ends) and the medial geniculate body (*c.g.m.*) have pushed themselves in. The inferior horn of the ventricle (*v.i.*) has opened out and one easily recognizes the condition of it produced by the rolling in of the hippocampal gyrus, from the free border of which the fimbria (*f*) passes upward. Laterally from the lateral geniculate bodies the optic radiation (*S*) passes through the posterior limb of the internal capsule toward the cortex of the occipital lobes. From the vicinity of the subthalamic region the most important part of the tegmental area has its origin, the subthalamic body has disappeared, the fillet (*S*) has broadened out, the red nucleus (*n.r.*) has become larger. The third ventricle passes beneath the posterior commissure (*c.p.*) into the aqueduct of Sylvius (*A.S.*). The medial peduncular fibres are covered over by the most anterior of the transverse pontal fibres (*P*). Posterior insular convolution, lenticular nucleus disappeared; pineal gland (*e*).

FIG. 2.—*Section through the Parietal Lobe* (posterior portion).

The brain stem with the pulvinar has completely separated itself from the hemisphere. The picture is similar to the first cross-section through the frontal lobe.

Corpus callosum (*c.c.*), posterior horn of the lateral ventricle (*v.p.*), association bundle (*Fli.*, *Farc.*), the optic radiation (*G.O.*) passes from the primary optic centres (corp. genic. lat., corp. quadrigem. anterior, pulvinar) farther in the occipital lobes.

Parietal convolutions, gyrus angularis (*g.ang.*), calcarine fissure (*f.c.*), supramarginal gyrus (*g.smg.*), lingula (*L.l.*), cuneus (*cun*).

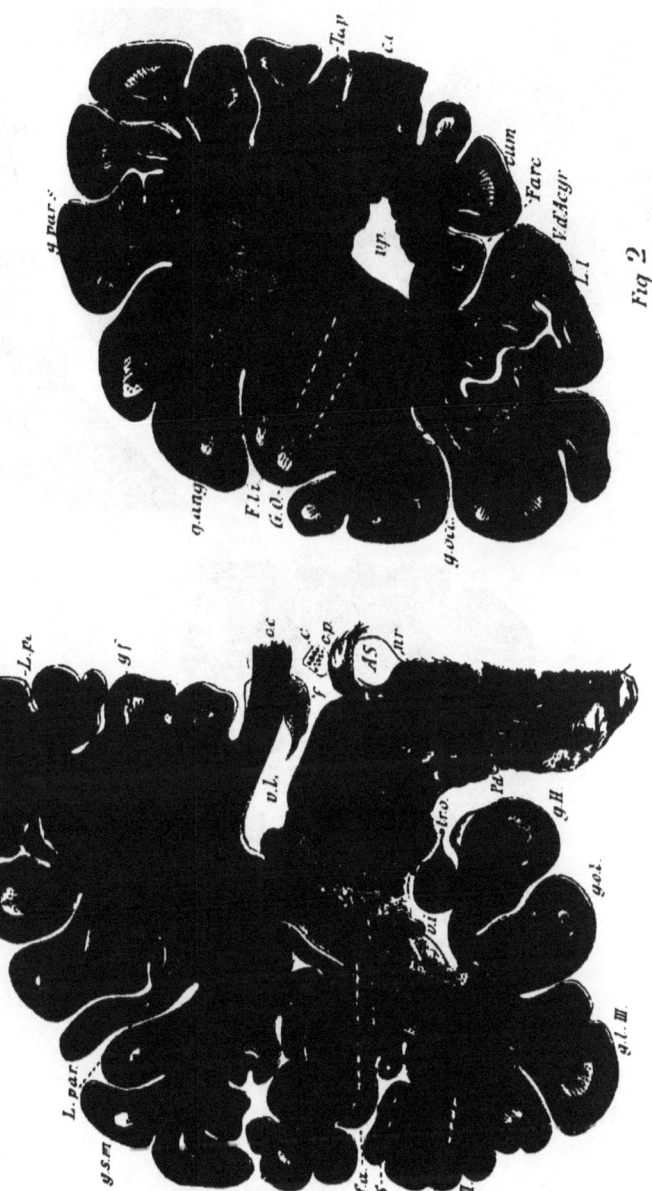

Tab. 28.

Fig 2

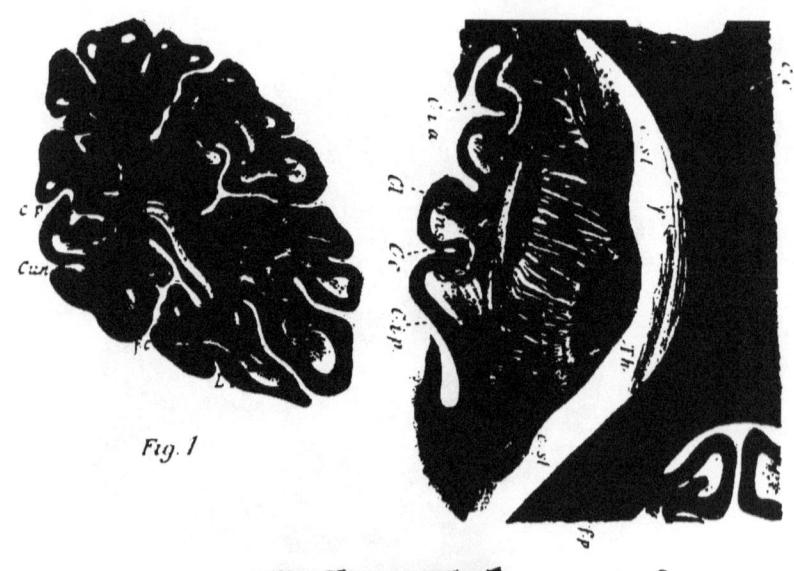

Fig. 1

Fig. 2

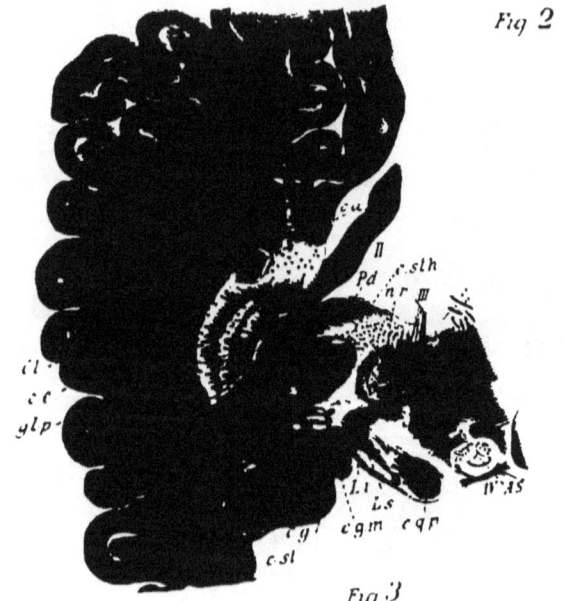

Fig. 3

Explanation of Plate 29.

Fig. 1.—*Frontal Section through the Posterior Pole of the Occipital Lobe.*
In the cortex of the cuneus, Vic d'Acyr's bundle. The optic radiation passes entirely into the cortex. Calcarine fissure ($f.c.$), lingual lobe ($L.l.$), cuneus (Cun), Occipital convolutions.

Fig. 2.—*Horizontal Section through the Superfices of the Corpus Striatum and the Optic Thalamus* (somewhat higher than in Plate 3, right half).
The internal capsule lies with broad bundles laterally to the cross section of the nucleus caudatus ($n.c.$) and thalamus opticus (Th). Somewhat externally is the lenticular nucleus (L), claustrum (Cl), and cortex of the island. Medially the lateral ventricle ($v.l.$) is opened, the fornix and corpus callosum ($C.c.$) are cut through. The anterior and posterior radiations of the corpus callosum (forceps anterior and posterior) are to be seen.

Fig. 3.—*Horizontal Section through the Cerebral Peduncle* (somewhat deeper than the section in Plate 4, right half). The section illustrates well the position of the internal capsule which is soon to pass into the pes pedunculi (Pd) with the lenticular nucleus ($L.,gl.p.$). The subthalamic region and the posterior corpora quadrigemina ($c.qp.$) are cut through. Laterally from the corpora quadrigemina the lateral geniculate bodies are included ($cg.l.$ and m). Optic tract (II), anterior commissure ($c.a$), trochlear nerve ($N.IV$), fillet, superior (Ls), inferior (Li), red nucleus ($n.r.$), subthalamic body ($c.sth$), substantia nigra ($S.n.$), motor oculi (III).

Explanation of Plate 30.

HORIZONTAL SECTION THROUGH THE ENTIRE LEFT HEMISPHERE IN THE MIDDLE OF THE CENTRAL GANGLIA.

The internal capsule (*c.i.*) is included in its entire extent; one sees easily the anterior (*c.i.a.*) and the posterior (*c.i.p.*) limbs, the knee, and behind, the optic radiations of Gratiolet (*G*).

Medially to it the caudate nucleus (*n.c.*) and optic thalamus, laterally the lenticular nucleus (*Put.*), capsula externa (*c.e.*), claustrum (*cl*). Tail of the nucleus caudatus in the posterior horn (*c.st.*).

Frontal lobes (*l.fr.*), central convolutions (*g.c.*), island of Reil (fossa of Sylvius, *f.S.*), temporo-occipital lobes (*l.p.* and *l.o.*).

The corpus callosum (*c.c.*) is cut through anteriorly and posteriorly, between which lie the septum pellucidum (*s.p.*), fornix (*f*), behind the choroid plexus (*pl. ch.*) covering the third ventricle, anterior horn (*c.a.*), inferior horn (*c.i.*) of the ventricle.

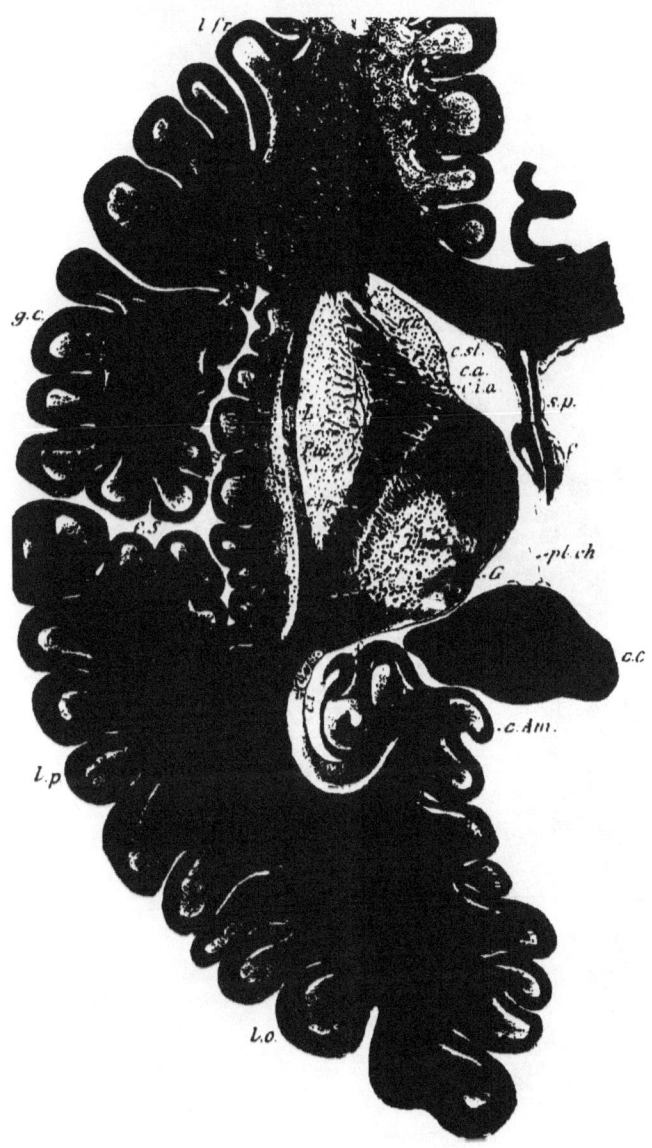

Tab. 31.

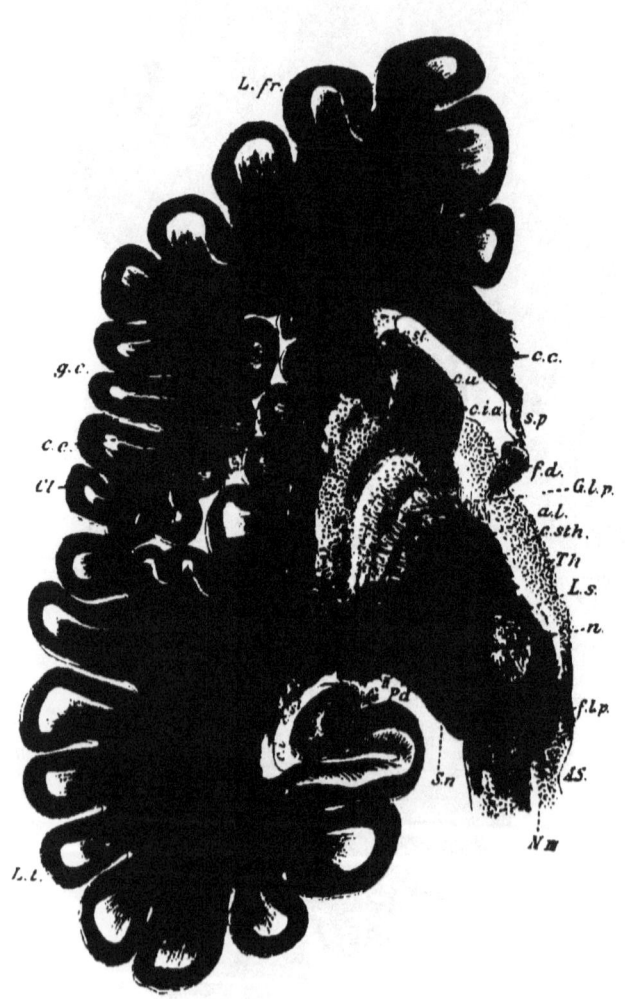

Explanation of Plate 31.

HORIZONTAL SECTION THROUGH THE BASE OF THE BRAIN STEM OF THE LEFT HEMISPHERE.

(About the same level as the section in Plate 4, right.)

Both limbs of the internal capsule are included in the section ($c.i.a$, anterior limb of the caspule); the posterior limb forms the principal portion of the crusta (Ped). The nucleus caudatus ($c.st$) and the base of the thalamus are cut across, in the subthalamic region the lenticular nucleus passes between the anterior fibres of the peduncle ($Pd.$) through the ansa lentiformis ($a.l.$), from the globus pallidus ($gl.p.$), and laterally therefrom its putamen (L). Subthalamic body ($c.sth.$). Substantia nigra ($S.n.$), lateral geniculate body (II), tail of caudate nucleus ($c.st$). Nuclear region of the motor oculi ($N.III$), fillet ($L.s.$), red nucleus ($n.r.$), anterior horn ($c.a.$), inferior horn ($c.i.$) of the lateral ventricle, frontal lobes ($L.fr.$), island, temporal lobes ($L.t.$), corpus callosum ($c.c.$), descending fornix ($f.d.$), septum pellucidum ($s.p$). Claustrum (Cl), external capsule ($c.e.$).

Explanation of Plate 32.

FIG. 1.—*Vertical Section through the Anterior Corpora Quadrigemina.*
(Closely following that of Plate 28, 1.)

The thalami (*Th.o.*) (pulvinar) are pressed back and outward from one another by the anterior corpora quadrigemina (*c.q.a.*) which press forward between them. The splenium corpus callosi (*c.c.*) with the fornix (*f*) lies over the anterior corpora quadrigemina.

In the central gray matter (*C.H.*) is the aqueduct of Sylvius (*A*), and beneath it the nucleus of the motor oculi nerve (*III*) (peripheral neuron), laterally from it the nasal [descending] root of the fifth (*Vn*).

In the corpora quadrigemina one differentiates a superficial and a deep medullary substance. Beneath these follow the tegmentum with the red nucleus (*n.r.*) of either side which gradually get closer to one another, laterally the superior (sensory) fillet (*Ls*), which includes externally the inferior (of the corpora quadrigemina) fillet (*Li*). Laterally to this the median geniculate body (*c.g.m.*). Separated from the tegmentum by the substantia nigra (*S.n.*) is the crusta (peduncle), and in its middle territory are the (motorial) pyramids. The optic tract (*Tr.o*) passes into the lateral geniculate body, the anterior [superior] quadrigeminal body, the pulvinar and there ends. From there the fibres of the optic radiation pass farther centrally (in the occipital lobes).

(Further details in photographic Plate 35, 2.)

FIG. 2.—*Section between the Anterior and Posterior Corpora Quadrigemina.*

The tegmental area has completely separated from the pulvinar thalami. The fillet (*L*) passes more caudally beneath the red nucleus. From the deep medullary substance (*st.i.*) the fountain-like tegmental radiation (*FF*) passes to the middle line where it undergoes decussation (raphe). Close beneath the motor oculi nucleus (*n.III*) in the central gray matter lies the posterior longitudinal bundle which is here very clearly recognizable (fasciculus long. post.) (*f*) laterally from which the thalamus fibres pass farther (substantia reticularis). In the motor oculi nucleus are readily recognized many individual nuclei. In the posterior brachium (*Br.a*) fibres pass from the posterior quadrigeminate body to the lateral geniculate body. In the substantia nigra fibres from the basal ganglion end.

Tab. 32.

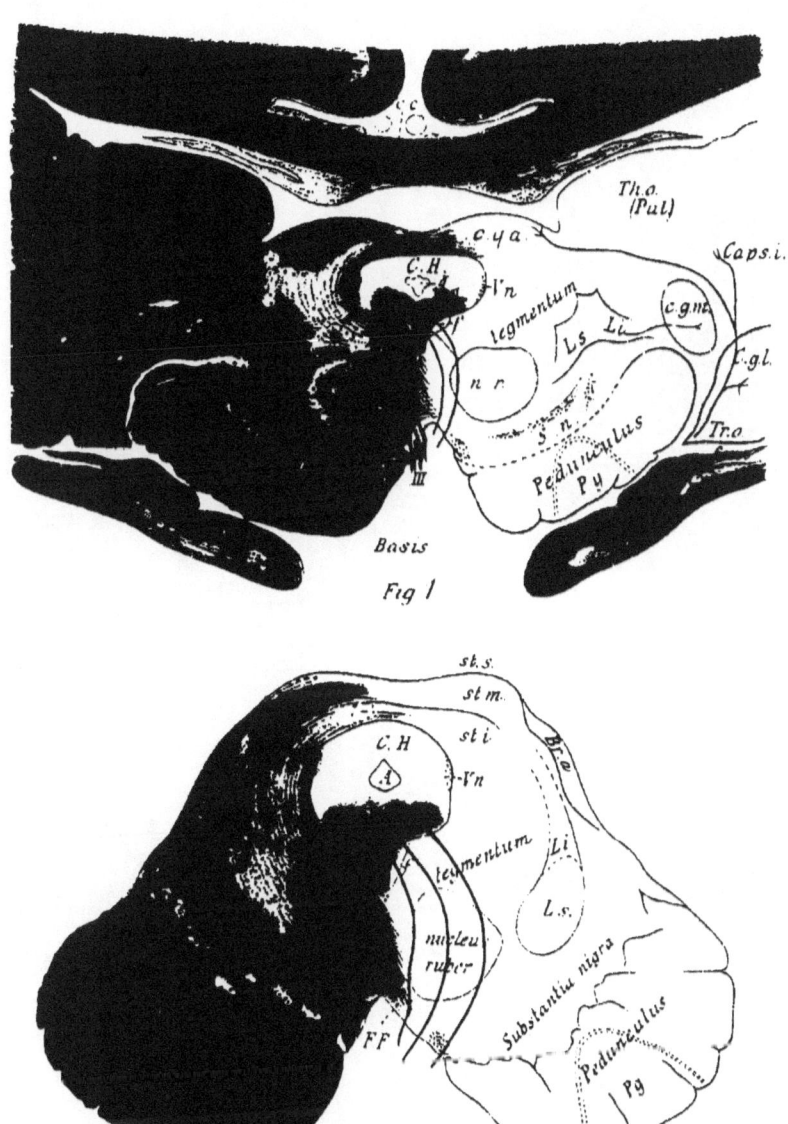

Fig 1

Tab. 33.

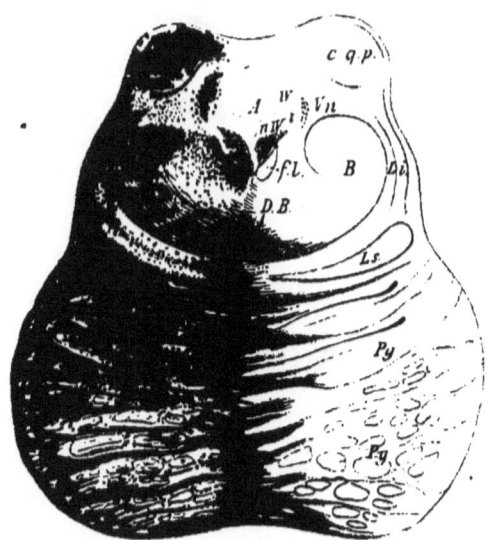

Fig. 1

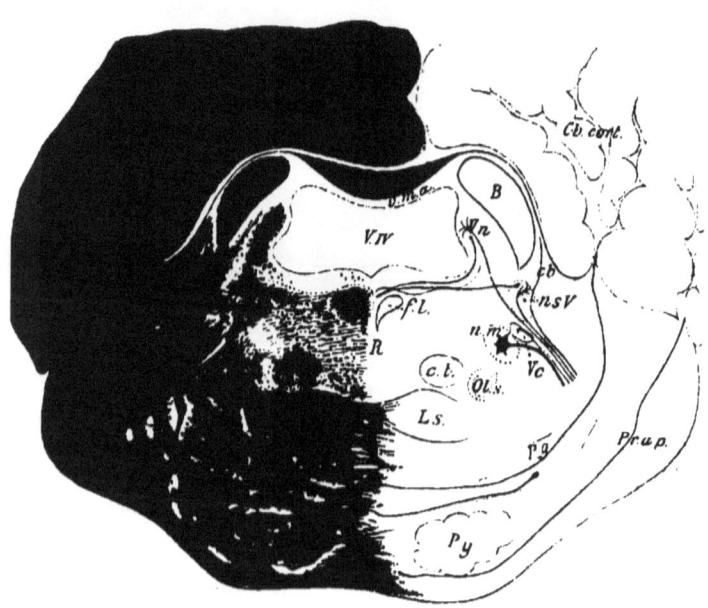

Fig 2

Explanation of Plate 33.

Fig. 1.—*Section through the Posterior Corpora Quadrigemina.*
The section includes basally the anterior portion of the pons which covers over the crusta. To the posterior corpora quadrigemina (*C.q.p*) passes a portion of the inferior (lateral) fillet (*Li*). Close to the nasal (descending) trigeminus root (*Vn*) the trunk of the trochlear nerve which originates here is involved (*IV*). In the tegmentum lie the fibres of the brachium conjunctivum, which originate in the red nucleus and decussate (*D.B.*) behind it (which pass to the cerebellum as the processus cerebelli ad corpora quadrigemina).

At the side of the posterior longitudinal bundle (*f*) is the reticulated substance of the tegmentum. Beneath the decussation of the brachium conjunctivum lies the horizontally placed upper (cortex-thalamus) fillet (*Ls*) (central sensory tract), and laterally to it the inferior (corpora quadrigemina) fillet (*Li*) (central acoustic tract).

One observes the enormous pontine ganglia (*Pg*) in which the larger part of the peduncular fibres (medial and lateral parts) end.

Fig. 2.—*Section through the Middle of the Pons.*
The aqueduct has widened into the fourth ventricle. Its roof is here formed by the velum medullare anticum (*v.m.a.*) with the lingula (*L*) (from the vermis of the cerebellum), laterally the enormous brachium conjunctivum (*B*) which have completed their decussation. In the tegmental territory are: the posterior longitudinal fasciculus (*f*), substantia reticularis tegmenti lateral from the raphe (*R*), central tegmental tract (*c.t.*), superior (medial) and inferior (lateral) fillet (*Ls* and *i*). Between the fibre tracts a number of cells are embedded (superior olive [*Ol.s*], nucleus of the substantia reticularis, etc.). Laterally from the tegmentum are the motor (*m*) and sensory (*s*) nuclei of the fifth and their roots to which the nasal root of the fifth (*Vn*) (in its vicinity the pigmented cells of the locus cœruleus) descends (motorial?); an additional tract follows from the neighboring medullary substance of the cerebellum as direct sensory cerebellar tract. Downward passes the caudal root of the fifth (*Vc*).

The pyramids (*Py*) pass (from the peduncles) covered over by the superficial and deep pontine fibres through the pons. From the cerebellar cortex comes the processus cerebelli ad pontem (*Pr.a.p.*).

Explanation of Plate 34.

FIG. 1.—*Section through the Posterior End of the Pons.*

The facial (*n. VII*) and the abducens nuclei (*n. VI*) have made their appearance in the tegmentum. The fibres of the facial (*VII*) pass as indicated on the right, arch around the abducens nucleus, collect at the knee (*g*), and then pass horizontally and eventually basalward out of the medulla (*VII*). The course of the abducens (*VI*) is more simple. Its nucleus is clearly in connection with the superior olive.

Substantia reticularis (*Srt*), fillet (*Ls*), etc., as before. Laterally from the trunk of the facial passes the caudal (descending) trigeminus root (*V.c.*) which has now made its appearance, with the sensory root of the fifth. To the outside of these the processus cerebelli ad pontem passes downward (*Pr.a.p.*), and between this and the brachium conjunctivum lie the restiform bodies (*s. C*) (process. cerebelli ad medull. oblong.) which can be seen more distinctly in the following section.

FIG. 2.—*Section through the Auditory Nucleus.*

In the medullary substance of the cerebellum is the corpus dentatum (*C.d.*), in the medullary substance of the vermis (*V*) and other lobes the nucleus tegmenti (*n.t*). Medially from the flocculus (*Fl*) of the cerebellum the acusticus (*VIII*) passes in. Its cochlear branch (*VIII.c*) ends in the ventral acoustic nucleus (*n. VIII.c.*) laterally to the restiform body (*c.r.*) (schematic on the right). Its vestibular branch (*n. VIII.v.*) ends in the dorsal acoustic nucleus (*n. VIII.d.*) and in the nucleus of Deiters just lateral to it (*n.D.*) (continuation of the locus cœruleus). From the central nucleus of the eighth the corpus trapezoides (*c.tr.*) passes transversely through the medial fillet (*L.s.*) (also to the superior olive) to the lateral fillet of the other side. Another portion of the (central) acoustic tract passes in the striæ acousticæ (*str.a.*) (schematic on right) to the same place. The pyramids (*Py*) have passed from out the pons.

(From here on the illustrations are photographs.)

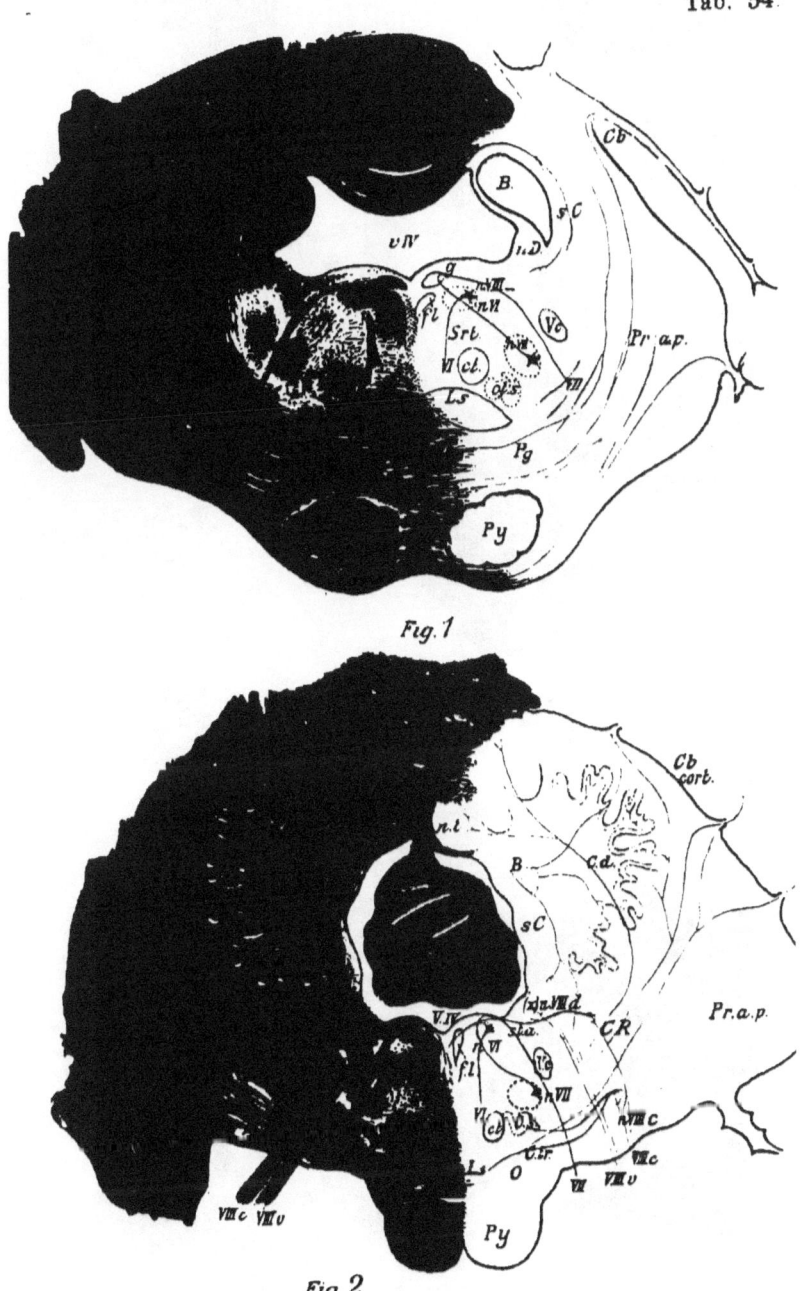

Tab. 34.

Fig. 1

Fig. 2

Tafel 35.

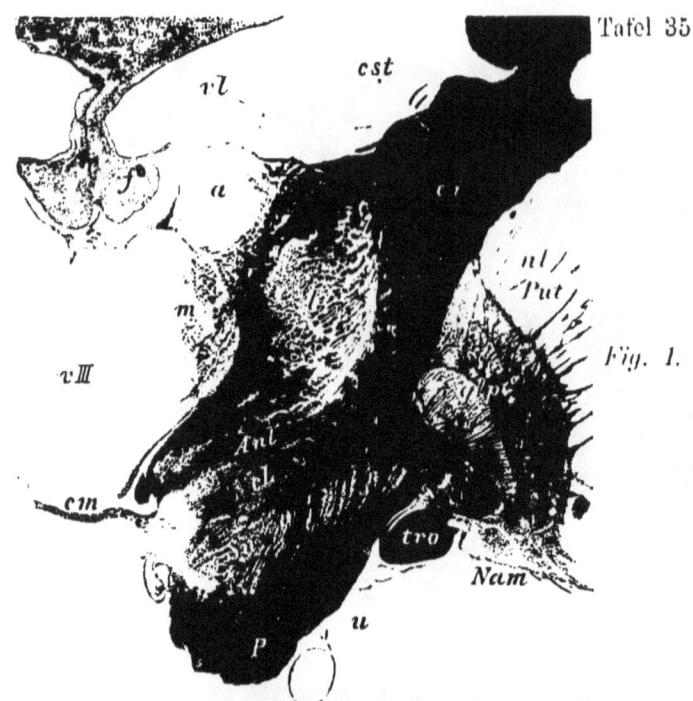

Fig. 1.

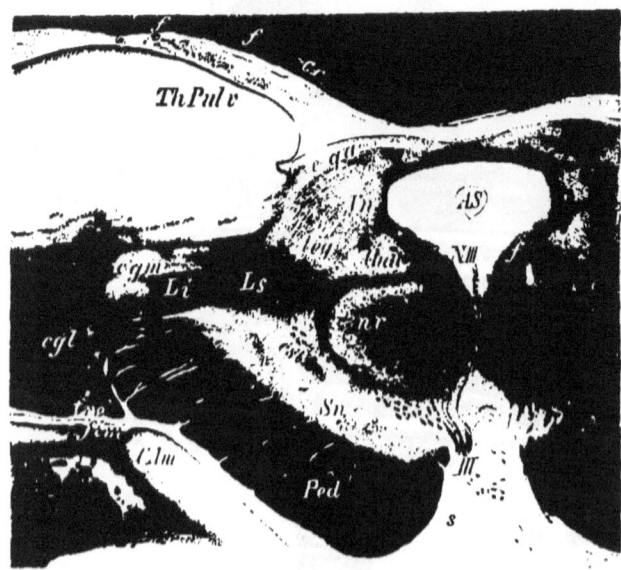

Fig. 2.

Explanation of Plate 35.

FIG. 1.—*Section through the Right Optic Thalamus at the Level of the Middle Commissure.*

(Sloping somewhat obliquely posteriorly.)

The three nuclei of the thalamus are easily recognized (nucleus anterior [*a*], medial [*m*], and lateral [*l*]). From the anterior there passes at this level a thick bundle (which in the horizontal section, Plate 31, is cut transversely [*v*] [Vic d'Azyr's bundle]) which ends in the corpus candicans of the same side (its course into the latter is well shown in the unstained section, Fig. 2, Plate 8). Beneath the thalamus lies the subthalamic region with the ansa lenticularis (*Anl*), Luys' body (*cL*), the fibres of which penetrate the internal capsule transversely and go to the globus pallidus (*glp*) of the lenticular nucleus; beneath the substantia nigra (*Sn*).

In the lateral nucleus (trellis layer) the corona radiata passes out of the internal capsule (*c i*) into the thalamus. The middle commissure (*cm*) contains but few nerve fibres. The medial nucleus forms posteriorly the pulvinar, in which ramify the fibres from the optic tract and the optic radiation (for other illustrations see Plate 27).

FIG. 2.—*Section through the Anterior Corpora Quadrigemina of the Left Side.*

One recognizes the more detailed abundance of fibres than in Plate 32, 1. The lettering is the same as in that figure.

In the lateral geniculate body (*cgl*) and also in the anterior corpora quadrigemina (*cqa*) end the innumerable arborizations of the fibres of the optic nerve. In this vicinity the tract for the pupillary reflex must be sought (from here to the oculomotorius nucleus, *NIII*). Beneath the multitude of outgoing motor oculi fibres (*III*) begins the crusta (*Ped*) in the medial segment of which can be seen Spitzka's bundle (*s*) (which contains probably the central tract for the motor cranial nerves) passing from the crusta up to the tegmentum and later decussating in the median line.

Explanation of Plate 36.

FIG. 1.—*Section through the Tegmentum behind the Posterior Corpora Quadrigemina.*

The aqueduct has already begun to expand into the fourth (vIV) ventricle.

For lettering see Fig. 2, Plate 33.

One observes especially the structure of the substantia reticularis (*Srt*) which conducts fibres from the optic thalamus downward and which contains ascending fibres from the anterior lateral columns of the spinal cord. Further than this very little is known. The median (upper) fillet (*Ls*) (principal fillet) is divided into several bundles; laterally is the inferior fillet (*Li*), which in part passes up to the posterior corpora quadrigemina.

Brachium conjunctivum (*B*), nasal root (*Vn*), nasal fasciculus of the trigeminus root (*Vn*), posterior longitudinal fasciculus (*f*); the innumerable fibre crossings in the raphe of the tegmentum is clearly represented (the central decussating sensory and the motor tracts pass here), locus cœruleus (*l.c.*), with strongly pigmented cells.

FIG. 2.—*Section through the Region of the Trigeminus Nucleus.*

The section contains practically the same structures as Fig. 2, Plate 33. The lettering is also the same.

One observes the finer details of the tegmentum and of the nucleus of the fifth. The caudal (descending) root with its bundles (*Vc*) is to be seen in this section very clearly between the motor (*m*) and sensory (*s*) nucleus. Central tegmental tract (*rt*), reticulated substance of the tegment (*Srt*); pons ganglia (*g*) between the superficial (*s*) and the deep (*p*) fibres of the pons (*P*). The raphe of the tegmentum (*K*) and the fibres of the pons contain numerous decussating pathways.

Below the superior olive (*os*) fibres (*Li*) of the corpus trapezoides and the inferior fillet (central auditory tract).

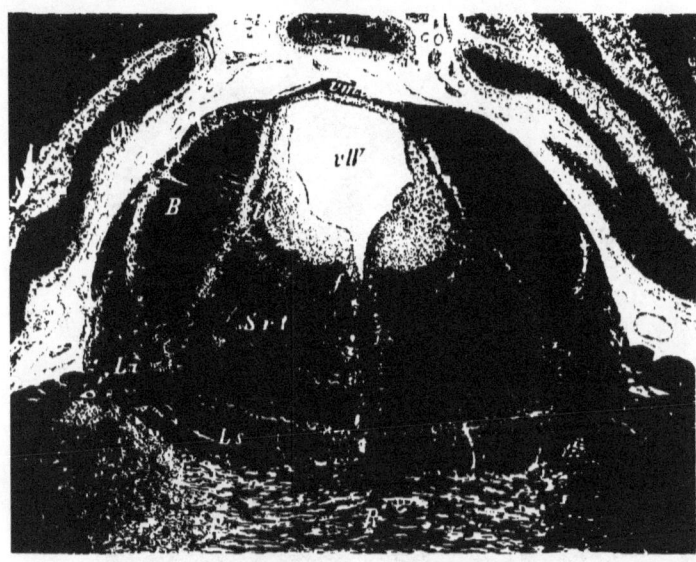

Fig. 1.

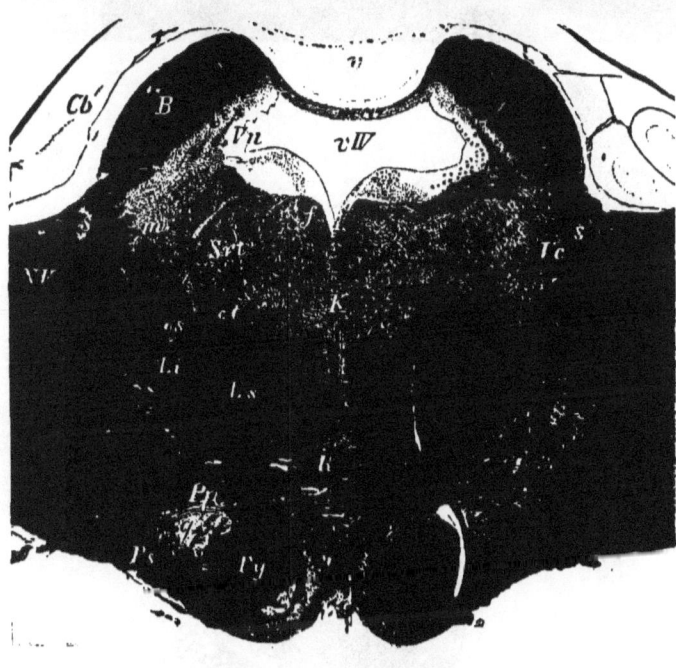

Fig. 2.

Taf. 37.

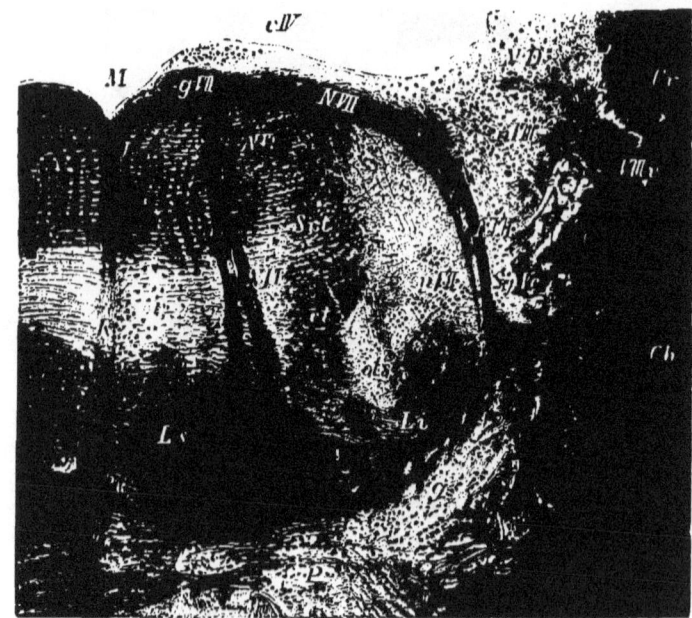

Fig. 1.

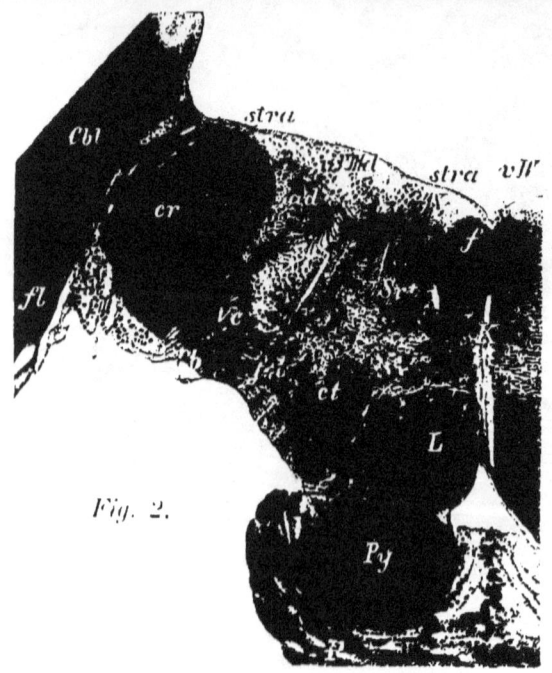

Fig. 2.

Explanation of Plate 37.

FIG. 1.—*Section through the Right Tegmental Region at the Level of the Facial Nucleus.*
One should compare the lettering with Fig. 1, Plate 34. From the deeply lying facial nucleus (*n. VII*) the individual fibres pass upward.

Likewise the pathway from the superior olive (*ol.s.*) to the abducens nucleus (*NVI*, white), (which last was more clearly recognizable somewhat further forward) is to be seen. From the cerebellum pass the restiform body (*Cr*) and the processus cerebelli ad pontem (*Cb*).

From the genu facialis (*gVII*) passes the united trunk of the seventh over the ependyma of the fourth ventricle (*vIV*). To the fillet passes from the neighboring ventral acoustic nucleus the corpus trapezoides (through the inner fillet).

Lateral from the trunk of the facial lies the descending root of the fifth (*Vc*), immediately in front of the substantia gelatinosa (*Sg*), which is the continuation upward of the posterior horns of the spinal cord. Dorsally therefrom is the dorsal nucleus of the eighth (*nVIII*) and the nucleus of Deiters (*n.D.*); superior, inferior fillet (*Ls.i.*); substantia reticularis (*Srt*); central tegmental tract (*ct*); raphe (*K*).

FIG. 2.—*Section through the Ventral Acoustic Nucleus of the Left Side.*
The restiform body (*cr*) has passed entirely out of the cerebellum and helps to form the lateral wall of the fourth ventricle. Laterally from it lies the semicircular ventral auditory nucleus in which the cochlear nerve goes. This is more developed in Fig. 2, Plate 34. From its central tract the black-colored fibres of the striæ acusticæ (*stra*), which can be distinctly seen here, pass over the floor of the fourth ventricle to the median line. They cross on the raphe (*R*) and pass farther forward in the lateral fillet. Dorsal nucleus of the eighth (*nVIIId*), medially therefrom the nuclear region of the glosso-pharyngeal (*nIX*), facial nucleus (*nVII*), substantia reticularis (*Srt*), central tegmental tract (*ct*), median fillet (*L*), pyramid, posterior longitudinal fasciculus (*f*), flocculus of the cerebellum (*fl*), descending acousticus root (*ad*).

Explanation of Plate 38.

FIG. 1.—*Section through the Cerebellum and Medulla.*

(Posterior portion of the Fourth Ventricle.)

The fourth ventricle (vIV) is here closed laterally only by the meninges as the restiform bodies (cr) have entirely passed out of the cerebellar hemispheres. The roof of the ventricle is formed by the vermis (V), the superior vermis (vs), the inferior vermis (vi).

In the cerebellum the corpus dentatum (cd) and the nucleus tegmenti (nt). The borders of the hemisphere are cut away. Nodulus (n).

In the medulla one recognizes the pyramids, the olives, the restiform bodies, etc.

FIG. 2.—*Section through the Medulla at the Level of the Glosso-Pharyngeal—Vagus Nucleus.*

Laterally from the pyramids the inferior olives (ol) have made their appearance. From the restiform bodies (cr) pass the cerebellar fibres (fol), decussating to the olive of the other side. Between the olives near the raphe (K) is represented the "interolivary layer" of the (superior) fillet (L); dorsally to it passes the posterior longitudinal bundle (f). Between the posterior longitudinal bundle and the restiform body lies the substantia reticularis (Srt) of the tegmentum. Dorsally to it the ninth and tenth sensory nuclei. Laterally from it the dorsal nucleus of the eighth. In front of the restiform body the descending root of the ninth, the tenth, and the solitary bundle (s) pass downward; beneath these the descending root of the fifth (Vc); laterally from the substantia gelatinosa (Sg). The fibres of the ninth and tenth nerves pass near by the solitary bundle out of the medulla. Medially from these in the tegmentum lies a nucleus (nucleus ambiguus), a cell group which is represented as a part of the motor nucleus of the vagus (na) and which can be seen better in the following sections. Median olive (olm), posterior olive (olp).

Tafel 38.

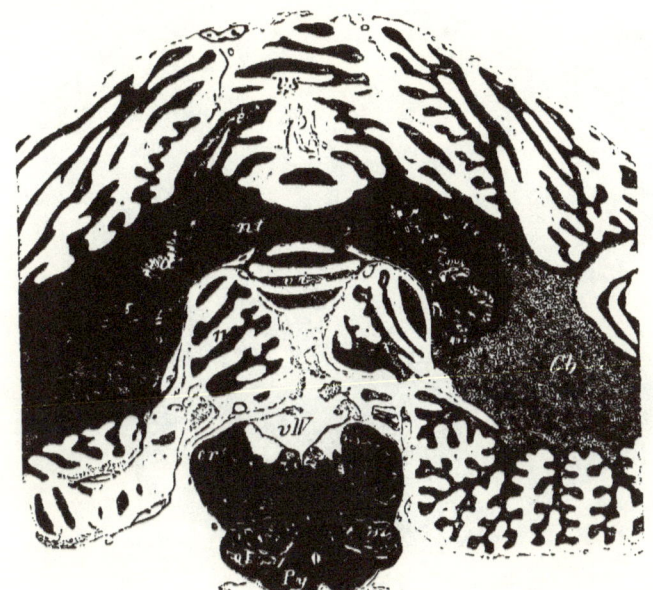

Fig. 1.

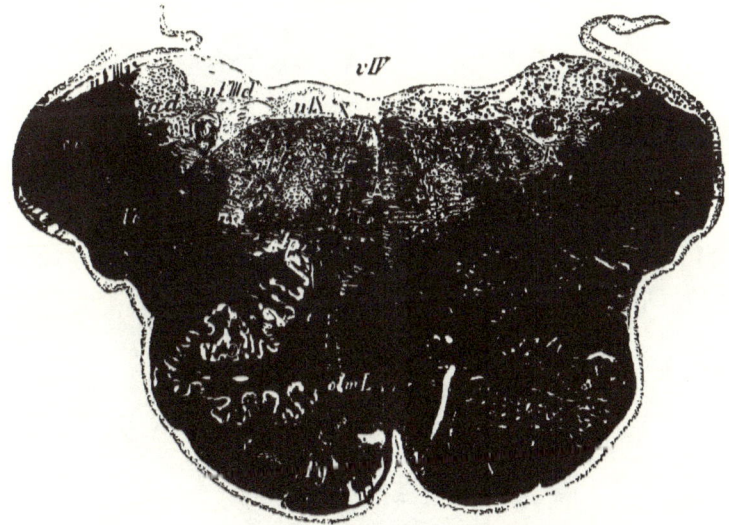

Fig. 2.

Tafel 39.

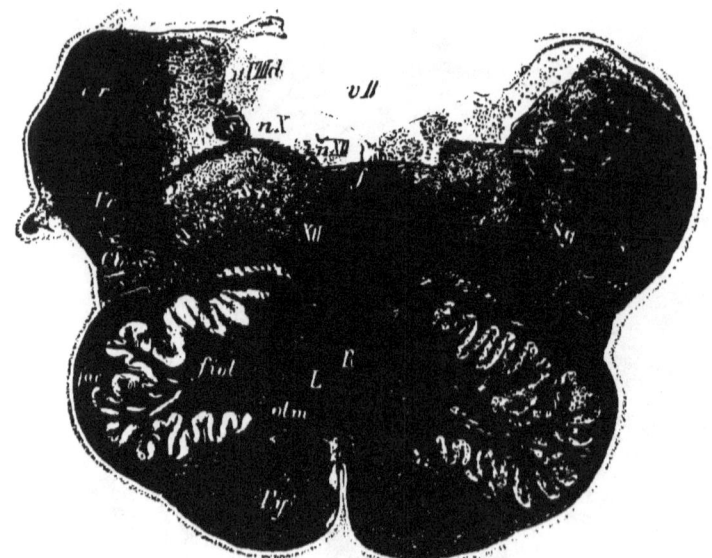

Fig. 1.

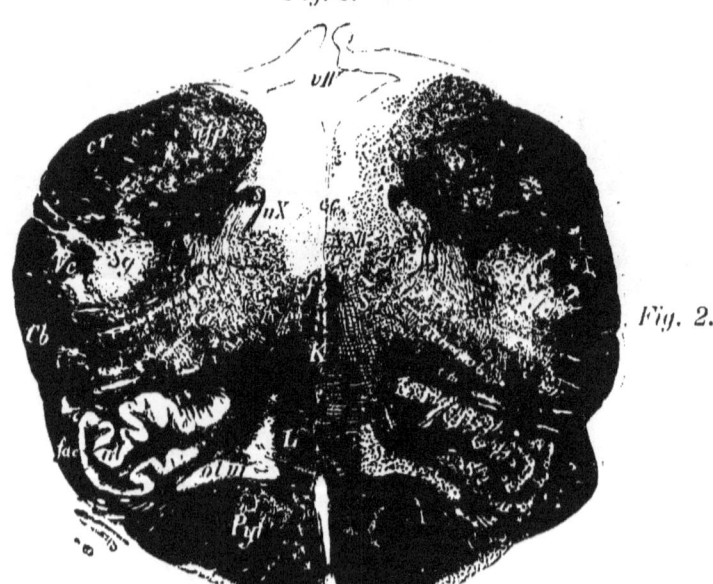

Fig. 2.

Explanation of Plate 39.

FIG. 1.—*Section through the Medulla at the Level of the Tenth and Twelfth Nuclei.*

Close to the nucleus of the tenth and medially to it passes the hypoglossal (XII) on the floor of the fourth ventricle, the fibres of which are penetrated externally by the posterior longitudinal bundle (f) and the fillet (L). Laterally to the olives pass the external arcuate fibres (fae) from the restiform body to the fillet of the same side, and through the substantia reticularis the internal arcuate fibres (fai) from the fillet to the most mesial portion of the restiform body of the opposite side (crossing therefore in the raphe). Above the inferior olive (ol) and mesially to it lies the inner accessory olive (olm). From the restiform bodies the lateral cerebellar tracts (Cb) pass off more deeply toward the pyramids. Everything else as in the previous section.

FIG. 2.—*Section through the Calamus Scriptorius of the Medulla.*

The restiform bodies (cr) approach one another, the fourth ventricle (vIV) becomes closed. The nuclei of the posterior columns pass into the restiform bodies, medial the nucleus of Goll, lateral the nucleus of Burdach (nfp). From these nuclei pass the massive internal arcuate fibres (fai) downward to the fillet (L) of the opposite side. The nucleus of the hypoglossus is more deeply situated ($nXII$). The nucleus of the tenth ends. Pyramids, fillet (L), substantia reticularis (Srt), substantia gelatinosa (Sg), descending root of the fifth (Ve), lateral cerebellar tract (Cb), solitary bundle (s), olives (ol), posterior longitudinal fasciculus (f), etc., unchanged. Raphe of the tegmentum (K).

Explanation of Plate 40.

FIG. 1.—*Section through the Nuclei of the Posterior Columns.*

Above the nuclei of Goll (nG) and Burdach (nB) the respective columns of these nuclei are to be seen distinctly (fG and fB).

The central canal with the nucleus of the twelfth ($NXII$) has sunken more deeply into the medulla.

The internal arcuate fibres (fai) and the external (fae) are plainly represented. All the other constituents as in the previous section, except that the olives (ol) have become considerably smaller.

FIG. 2.—*Section through the Medulla beneath the Olives.*

After the disappearance of the olives the medulla becomes considerably smaller.

The nuclei of the posterior columns are smaller (NG, NB), the bundles which arise from them (funiculus Goll and Burdach) larger. Near and lateral to the nucleus of the twelfth the nucleus of the ninth has appeared, the fibres of which pass out transversely through the medulla.

The lateral cerebellar tracts (Cb) are represented by a clearly differentiated area, dorsally from which is the substantia gelatinosa (Sg) and laterally the descending root of the trigeminus (Vc). Internal and external arcuate fibres, substantia reticularis (Srt), fillet (L) (which has become considerably smaller), fillet decussation (K), pyramids, inner olives.

Between Goll's columns the posterior longitudinal sulcus (Sp), between the pyramids the anterior longitudinal sulcus (Sa). In the pyramids the nucleus arciformis (na).

Tafel 40.

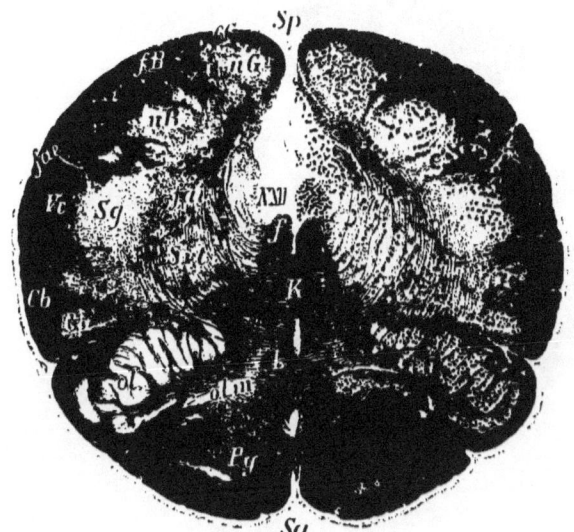

Fig. 1.

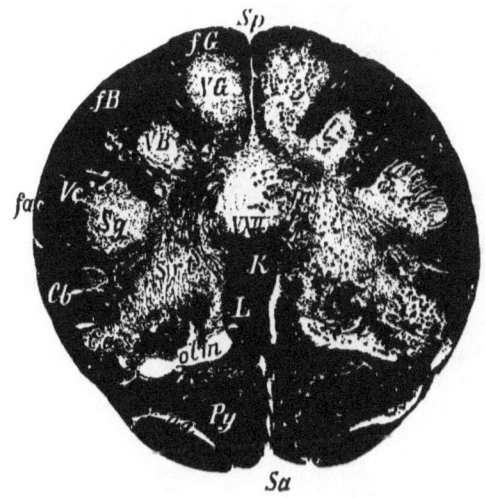

Fig. 2.

Tafel 41.

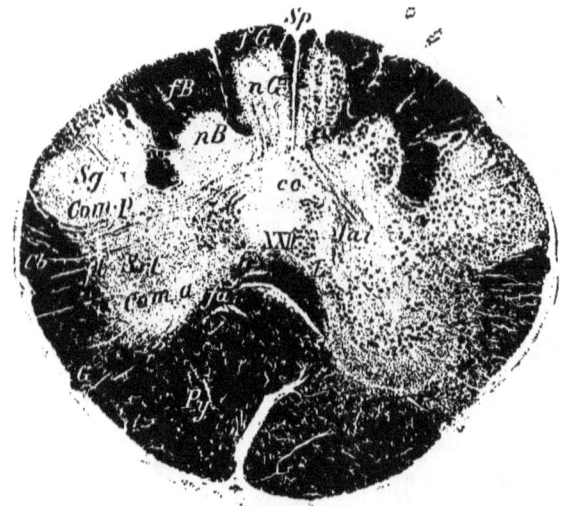

Fig. 1.

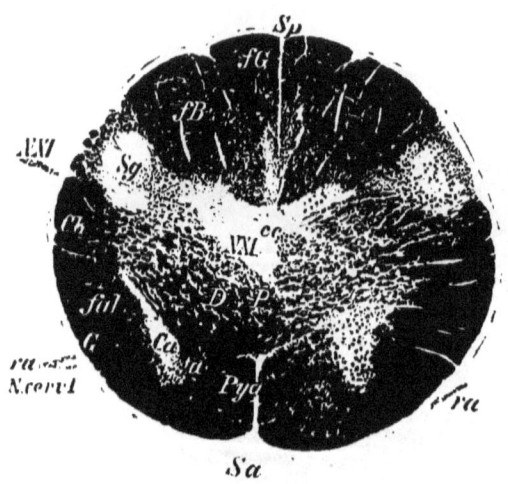

Fig. 2.

Explanation of Plate 41.

FIG. 1.—*Section of the Medulla Immediately Above the Pyramidal Decussation.*

The pyramids (Py) are more deeply situated in the medulla and displace one another and thus push asunder the small fillet layer (fa). At their apices the posterior longitudinal fasciculus (f) is still to be seen. The internal arcuate fibres (fai) pass distinctly out of the nucleus of Goll's column (nG) to the crossed fillet layer.

The nuclei of the posterior columns (nG, nB) are smaller, the substantia gelatinosa (Sg) gets continually larger (beginning of the posterior horns). In the substantia reticularis tegmenti (Srt) the area alongside of the remains of the fillet and the posterior longitudinal fasciculus is later to be the position of the anterior horns ($corn.a$). Laterally from this the substantia reticularis is displaced by the appearance of the antero-lateral ground bundle (fl) and the column of Gowers (G) and posteriorly the lateral cerebellar tract (Cl). In the middle the nucleus of the twelfth which has nearly disappeared and the nucleus of the eleventh which is still distinct. The accessory olives have disappeared. The spinal portion of the eleventh nerve arises from a group of ganglionic cells lying in the lateral segment of the anterior horn.

FIG. 2.—*Section through the Pyramidal Crossing.*

The nuclei of the posterior columns have disappeared, the territory is entirely taken up with the fibres of the posterior columns, Goll's and Burdach's columns (fG, fB).

The pyramids pass decussating, severing the anterior horns ($D\ Py$), deeply into the lateral columns of the opposite side.

The posterior horn (Sg) is distinctly recognizable. All the other parts as in the cut above.

Anterior horn (Ca), central canal (cc), sulcus anterior (Sa), sulcus posterior (Sp).

Explanation of Plate 42.

The sections from here on, in conformity with the usual custom, are represented reversed. The dorsal portion (Sp) downward and the ventral (Sa) upward.

FIG. 1.—*Section through the Cervical Cord Immediately Beneath the Pyramidal Decussation.*

The pyramids (Py) have in large part disappeared from the anterior columns and are found in the crossed pyramidal tracts. In the anterior columns remain the small uncrossed pyramidal tract (Pya) and the anterior ground bundle (fa) (from the posterior longitudinal fasciculus). In the lateral columns are found the crossed pyramidal tracts (Py); the antero-lateral ground bundle (fal), the lateral limiting layer (fl), and Gowers' bundle as a continuation of fibres from the substantia reticularis tegmenti; the lateral cerebellar tracts (Cb) from the restiform bodies (ventral segment).

In the posterior columns lie the columns of Goll (fG) and Burdach (fB).

At the bottom of the anterior longitudinal fissure lies the anterior commissure (d), the spinal prolongation of the pyramidal decussation, behind it the central canal ($c.c.$) and the posterior commissure.

The anterior (Ca) and posterior (cp) horns are completely developed and in the posterior horns the substantia gelatinosa (Sg), which was remarked above, has become prominent. The roots of the uppermost cervical nerve pass here out of the anterior horn (anterior root), and in the lateral segment of Burdach's column is seen the entering posterior root (in the posterior root zone).

FIG. 2.—*Section through the Upper Cervical Cord, Level of the Fourth Cervical Nerve.*

Lettering as in Fig. 1. The substantia gelatinosa (cp) of the posterior horn is much smaller, as it is in all the following sections. Origin of the phrenic in the anterior horn cells (Ca).

Tafel 42.

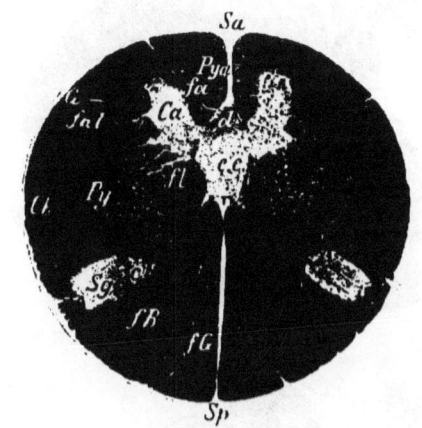

Fig. 1.

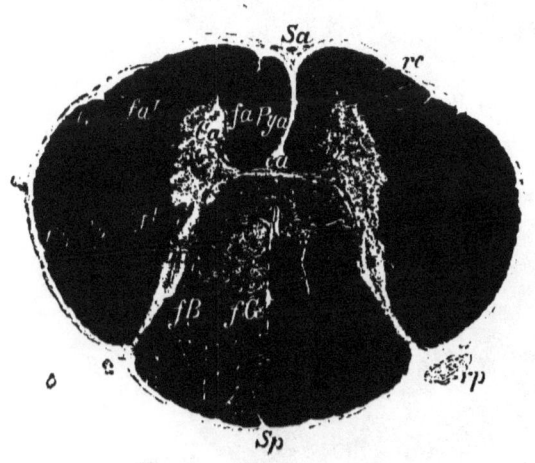

Fig. 2.

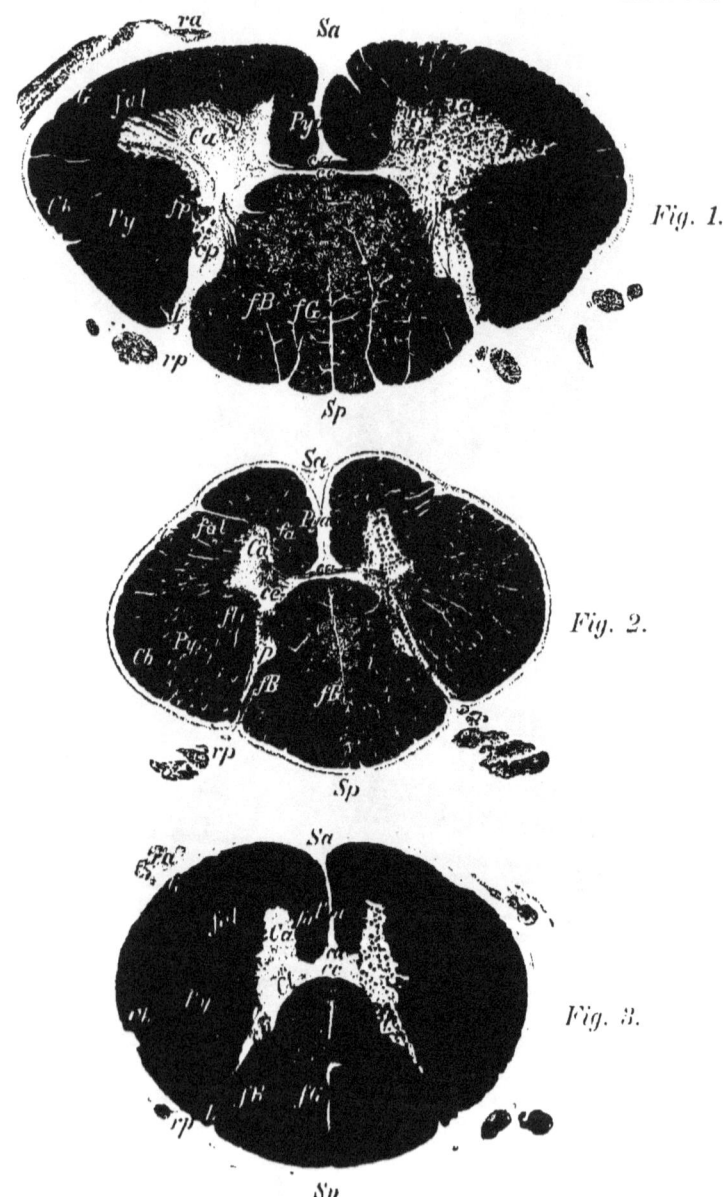

Tafel 43.

Fig. 1.

Fig. 2.

Fig. 3.

Explanation of Plate 43.

FIG. 1.—*Section through the Cervical Enlargement, Level of the Seventh Cervical Nerve.*

The gray substance on account of the enormous development of the anterior horns (*Ca*) is considerably increased. The composition is the same as given for Plate 42, 1. In the anterior horns are the cells for the peripheral motor neurons of the arm muscles (at this level especially for the forearm musculature). There are differentiated a mesal (small) (*m*) and a lateral (*l*) ganglion-cell groups. The lateral (and the mesal) divide further into an anterior and a posterior division (*la* and *lp*). Further concerning localization is not known positively. The lateral group may contain the proper motorial cells. Behind the motorial cell groups lie the so-called "middle cells" (*c*), around which ramify innumerable collaterals from the posterior roots and also from the lateral columns. The posterior roots (*rp*) and their course are seen better in Plate 47, 1.

FIG. 2.—*Section through the Upper Dorsal Cord, Level of Third Dorsal Nerve.*

FIG. 3.—*Section through Middle of Dorsal Cord, Level of Sixth Dorsal Nerve.*

The anterior horns have again become small and from them arise the peripheral neurons for the intercostal muscles. At the base of the posterior horns one can make out the rudiments of Clarke's columns (*Cl*). Central canal (*cr*).

Goll's columns get smaller the further down we go.
The remaining relationships are unaltered.
Anterior roots (*ra*), posterior roots (*rp*).

Explanation of Plate 44.

FIG. 1.—*Section through the Lower Dorsal Cord, Level of the Eleventh Dorsal Nerve.*

The anterior horns (Ca) increase in size and their configuration becomes changed; Clarke's columns with their cells are large and sharply defined. From them fibres pass to the lateral cerebellar tract of the same side (Cl) and pass with this tract to the crossed nucleus of the tegmentum in the restiform body. The lateral pyramidal tracts (Py) pass, below the starting-point of the lateral cerebellar tract, quite out to the periphery. The anterior pyramidal tract becomes smaller and is lost in the upper portion of the lumbar cord. All the other columns become smaller because they have given off the greater portion of their fibres in segments higher up.

FIG. 2.—*Section through the Upper Lumbar Cord, Level of Second Lumbar Nerve.*

FIG. 3.—*Section through the Lower Lumbar Cord, Level of Fourth Lumbar Nerve.*

The anterior and posterior horns become very much extended. This is due principally to the increase of the gray substance. The cell groups in the anterior horns are similar to those in the cervical region. From them arise the peripheral neurons for the leg muscles. The posterior roots (rp) enter in thick bundles the posterior root zones, and their collaterals and short branches radiate bow form into the posterior horns.

For further details see Plate 47, 2.

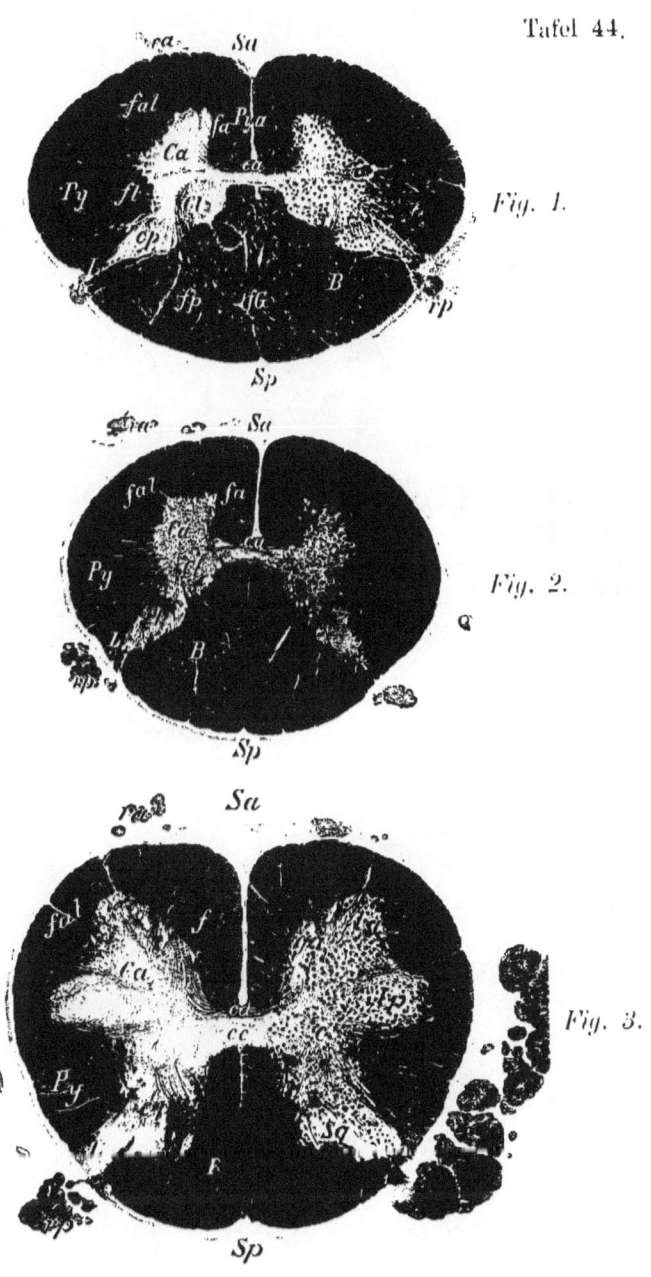

Tafel 44.

Fig. 1.

Fig. 2.

Fig. 3.

Tafel 45.

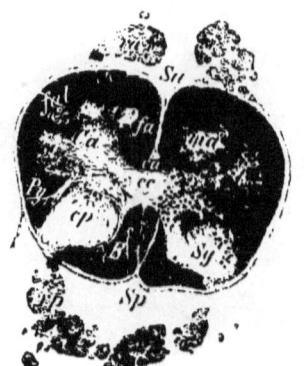

Fig. 1.

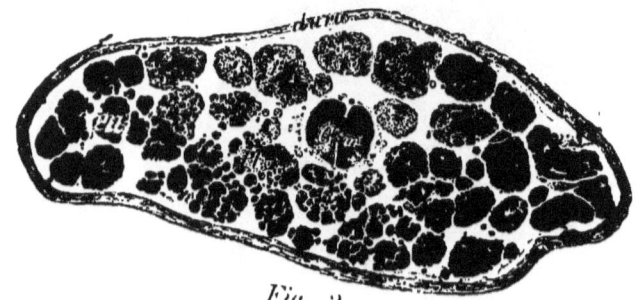

Fig. 2.

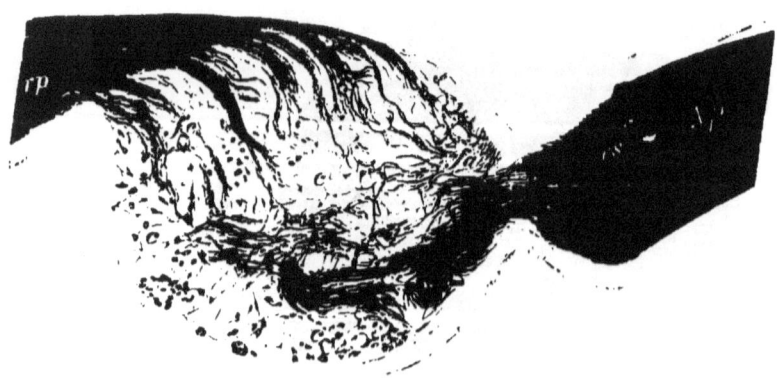

Fig. 3.

Explanation of Plate 45.

Fig. 1.—*Section through the Sacral Cord, Level of Third Sacral Nerve.*

The pyramidal columns terminate in the small white substance of the lateral columns. The gray substance is relatively greater than the white. In the broad anterior horns are situated the motor ganglion cells that give origin to the peripheral neurons for the small muscles of the feet, somewhat lower are those for the anal and vesical muscles and their reflex automatic activity.

Fig. 2.—*Section through the Cauda Equina and the Conus Medullaris.*

In a sac of the dura lie the more deeply passing motor and sensory roots, given off from the spinal cord in the lumbar region, for the lower extremities. Laterally the roots which later form the crural (*Pll*), medially, the sciatic continuation (*Pls*). The first, after their exit, form the lumbar plexus, the other the sacral plexus. In the centre lies the inferior termination of the spinal cord, the conus medullaris (*C m*).

Fig. 3.—*Section through a Posterior Root and Spinal Ganglion from the Lumbar Cord.*

The mixed nerve (*Np*) coming from the periphery goes as far as the spinal ganglion, its sensory portion (*rs*) passes into the ganglion and ends in the cells of the latter (one can see despite the weak magnification the strongly pigmented cells of the ganglion). From it passes collectively the posterior root (*rp*) out of the ganglion and into the spinal cord (especially into the posterior root zone of the posterior column).

Explanation of Plate 46.

FIG. 1.—*Section through the Sciatic Nerve at its Upper Point of Emergence.*

(Great sciatic foramen.)

Enveloped by the epineurium (*P*), with weak magnification one recognizes the numerous collected individual bundles (*f*) of varying thicknesses. Between them the blood-vessels course in the perineural connective tissue.

FIG. 2.—*Longitudinal Section of a Nerve Bundle taken from the Sciatic Nerve.*
For explanation see Fig. 3.

FIG. 3.—*Transverse Section of a Nerve Bundle from the Sciatic Nerve.*

One recognizes with strong magnification the segregation of medullated fibres (stained black) of different thickness. In all the nerve bundles there course without order, close to the thick fibres, numerous very small nerve fibres surrounded by a scarcely recognizable medullary sheath. In the nerve bundles individual compartments are formed by the endoneural connective tissue.

FIG. 4.—*Cross Section through a Normal Optic Nerve* (weak magnification), in its nerve sheath formed from the dura (*v*). It consists also of numerous individual bundles in which the nerve fibres of very fine calibre course.

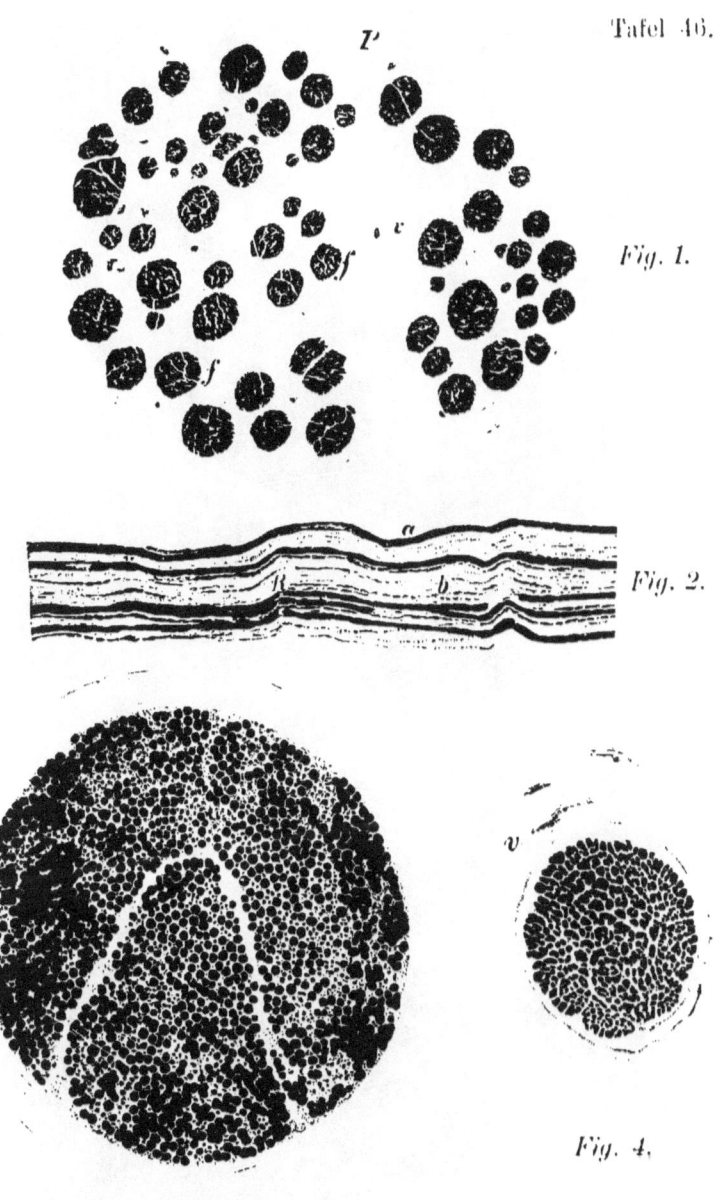

Tafel 46.

Fig. 1.

Fig. 2.

Fig. 3.

Fig. 4.

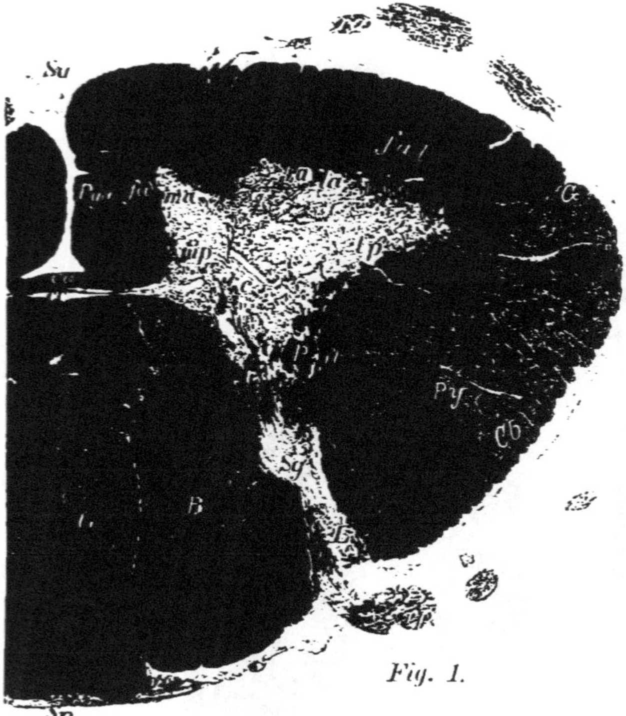
Su
fal
la
G
fa
ma
mp
lp
c
P
Py
Cb
Sg
B
L
Sp

Fig. 1.

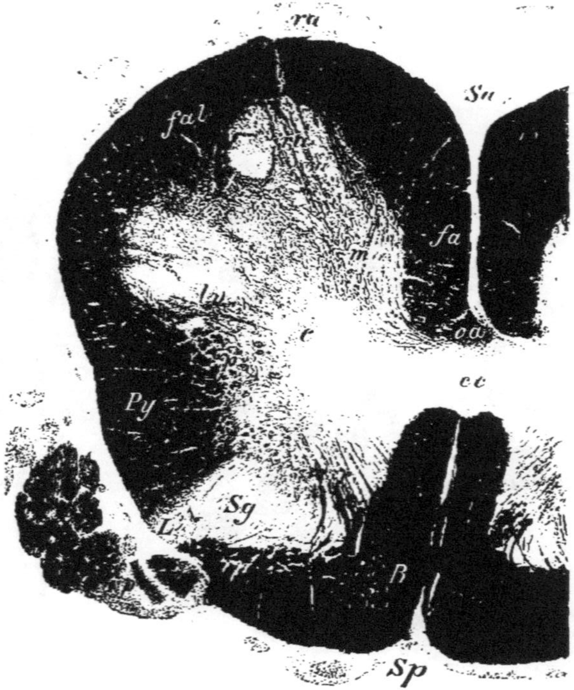
ra
Su
fal
fa
m
la
c
oa
cc
Py
Sg
L
B
Sp

Fig. 2.

Explanation of Plate 47.

THE GRAY SUBSTANCE OF THE SPINAL CORD.

The two photographs represent with strong magnification the right half of a cervical section and the left half of a lumbar section (comp. Plate 43, 1 and Plate 44, 3).

One observes particularly the enormous number of fibres going in and coming from the horns. In the anterior horn there pass in:

1. Fibres from the anterior commissure (decussating anterior pyramidal columns and decussating central sensory fibres).

2. Fibres from the antero-lateral tract (motorial end fibres and collaterals from the lateral pyramidal tract and central sensory collaterals from the anterior lateral columns).

3. Fibres from the posterior horn (reflex collaterals and other posterior root fibres).

From the anterior horn pass out the anterior (motor) roots arising from the motor ganglion-cell groups; they pass out in bundles through the antero-lateral tracts. (They arise especially from the lateral cell groups.)

Into the posterior horn there passes the thick bow-formed radiations of the collaterals and the collected short tracts of the posterior roots from the posterior column.

At the posterior end of the posterior horn formed by the substantia gelatinosa (Sg) lie the fields of Lissauer (medullary bridge), consisting of thin posterior root fibres (L), and immediately in front of them the zona spongiosa of the posterior horns.

Explanation of Plate 48.

MEDULLATION OF THE FŒTAL BRAIN (photographs).

This plate follows immediately after Plate 11 and is used for the first time now on account of being more readily comprehended.

FIGS. 1 to 5 represent the unstained sections from the brain stem of a fœtus eight months old. By far the greater number of fibres which in the fully developed brain are medullated are here unmedullated (therefore unstained). Nevertheless there are individual phylogenetically manifest old and important tracts which are on account of their medullation easily recognized macroscopically. They appear brilliantly white in contrast with the remaining gray matter of the brain stem. The most important of these tracts are: the medullary radiation to the red (nr) nucleus (Fig. 1, L), the ansa lenticularis and the fillet (Ls, Li). In addition to this there are also medullated the posterior longitudinal bundle, the nasal (descending) root of the trigeminus, the peripheral cranial nerves, the optic tract, the posterior commissure, and others.

FIGS. 1 AND 2.—*Section through the Middle and Posterior Segment of the Optic Thalamus.* One observes that the crusta (motor pyramidal tract, etc.) is not yet medullated.

FIG. 3. *Section through the Anterior Corpora Quadrigemina.*

FIG. 4.—*Section through the Middle of the Pons.*

FIG. 5.—*Section through the Medulla Oblongata.*

The fillet passes in the first section at the height mentioned above almost entirely as a portion of the ansa lenticularis and further through the globus pallidus and (probably not interrupted) in the so-called tegmental radiation (the first medullated pathway of the cerebral hemispheres, sensory tract! See text, Section IV., p. 65) to the cortex of the posterior central convolutions and the parietal lobes.

FIGS. 6 and 7 are sections through the cervical cord and dorsal cord of a new-born. One recognizes the not yet medullated pyramidal tract (Py) sharply differentiated from the fully developed lateral cerebellar tract (Cb). For description of the dorsal cord see Plate 10, Fig. 2 (medullary sheath staining).

Tafel 48.

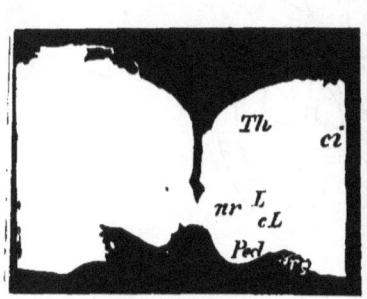

Fig. 1.

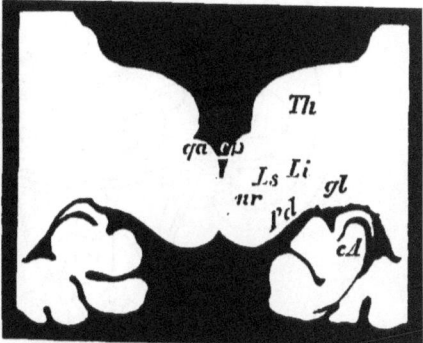

Fig. 2.

Fig. 3.

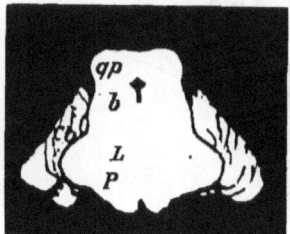

Fig. 4.

Fig. 5.

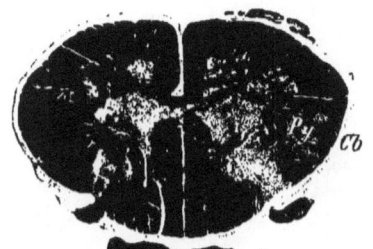

Fig. 6.

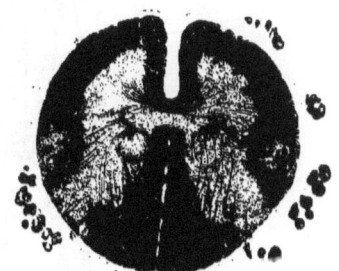

Fig. 7.

Tab. 49.

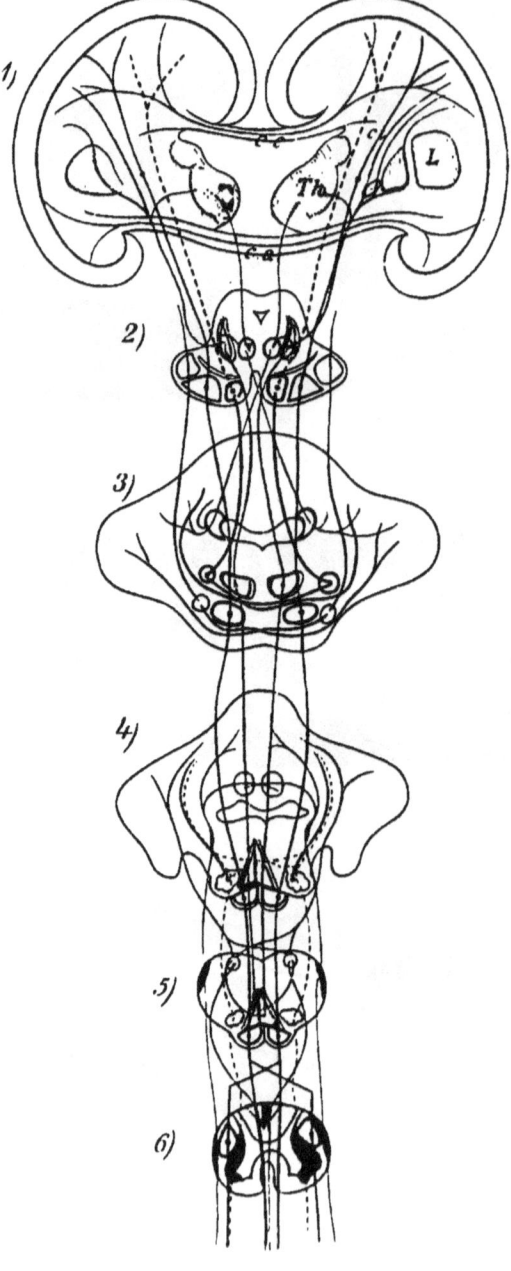

Explanation of Plate 49.
SCHEME OF THE COURSE OF THE MORE IMPORTANT CEREBRAL TRACTS.

In the scheme, which is comprehensible without further description, there are represented (1) transverse sections of the hemispheres, (2) the cerebral peduncles, (3) the pons, (4) the upper, (5) the lower medulla, and (6) the spinal cord. Their respective positions are clearly discernible in the plate.

1. Blue stippling, the frontal pons-cerebellar tract, from the pons ganglia to the opposite cerebellar hemisphere (1–3).

2. Yellow, the temporo-occipital pons-cerebellar tract, from the pons ganglia to the opposite cerebellar hemisphere (1–3).

3. Red, the motorial pyramidal tract (central neuron) decussates almost completely between 5 and 6 and passes in the cord, partly in the lateral column, partly in the anterior column to the motor nuclei of the medulla in the brain and to the motor nuclei of the anterior horns in the cord.

4. Green, the sensory fillet tract (central neuron), the important portion of the central tegmental tract (ansa lenticularis). It breaks through the internal capsule, passes farther in the tegmentum as superior fillet, and in 5 (internal arcuate fibres) decussates to the nuclei of the posterior columns. The greater portion is certainly interrupted somehow in the thalamus or in the subthalamic region.

5. From the thalamus passes a tract (brown) to the red nucleus, from there (2) the brachium conjunctivum decussates to the cerebellar hemisphere.

6. The lateral cerebellar tract (brown 4–6) passes from the spinal cord (Clarke's column) to the restiform bodies and so to the vermis of the cerebellum (crossed tegmental nucleus).

7. From the nuclei of the posterior columns on both sides fibres pass (green) to the cerebellum as external arcuate fibres to the restiform body (4–5).

8. From the olive (4) fibres (stippled yellow) of the opposite side pass to the cerebellum, olivary fibres of the restiform body.

9. In cross section of the cerebrum (1) the commissural tracts (brown) pass in the corpus callosum and the anterior commissure from the cortex of one side to the other.

Explanation of Plate 50.

Figs. 1 and 2 show the relationship of the tegmentum and the foot of the cerebral peduncle to the cerebral hemisphere tracts.

Fig. 1 shows the hemisphere on frontal section, Fig. 2 on horizontal section.

In Fig. 1 are seen on the left the connections of the crusta, on the right the connections of the tegment.

Crusta: frontal pons tract (blue), pyramidal tract (red), temporo- (parieto-) occipital pons tract (blue stippling).

Tegment: fillet tract (green), thalamus—red nucleus (brown).

In the section of the hemispheres there are represented schematically the long and short association tracts (brown), and the corona radiata to the thalamus (superior and inferior stalks) (brown).

In Fig. 2 in addition, the optic radiations (occipital cortex—optic centres—optic tract) (yellow).

Fig. 3.—*Scheme of the Optic and Motor Oculi Tracts and their Connections.* (The fibres of the left optic tract are colored.)

Optic: peripheral neuron (brown) in the retina ($b=bulb$); central neuron (green), in the optic nerve to the chiasm (decussation of the nasal bundle); from there in the optic tract to the lateral geniculate body, pulvinar, anterior quadrigeminate body. Here the fibre arborization occurs.

From there the fibres pass in the optic radiations to the cortex of the cuneus of the occipital lobes (it is doubtful if there exist direct fibres from the tract to the cortex collaterals?).

Motor oculi: central neuron (red) from the cortical centre (angular gyrus?) crossed to the motor oculi nucleus on the floor of the aqueduct.

Peripheral neuron (blue), from there as peripheral nerve (of crossed origin also, for the rectus internus) to the musculature of the bulb.

Pupillary reflex tract: optic fibres—corpora quadrigemina—reflex collaterals (stippled yellow, where?)—motor oculi nucleus—peripheral nerve (blue)—sphincter pupillæ (intercalated in ciliary ganglion).

Tab. 50.

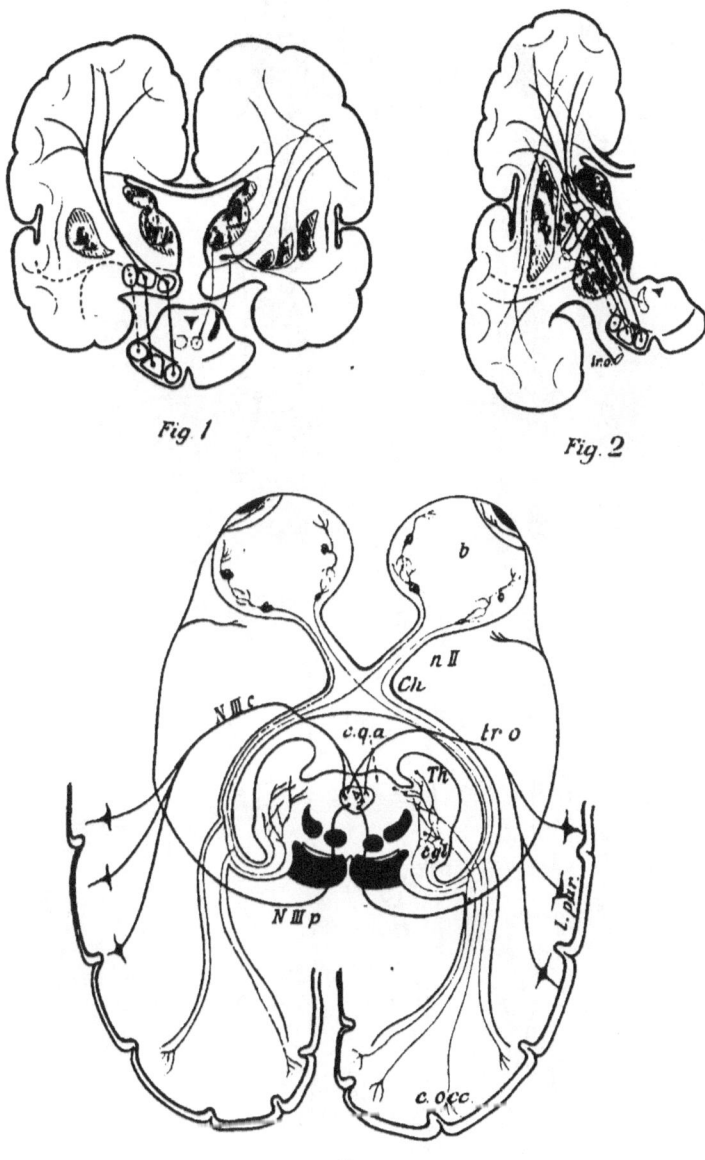

Fig. 1

Fig. 2

Fig. 3

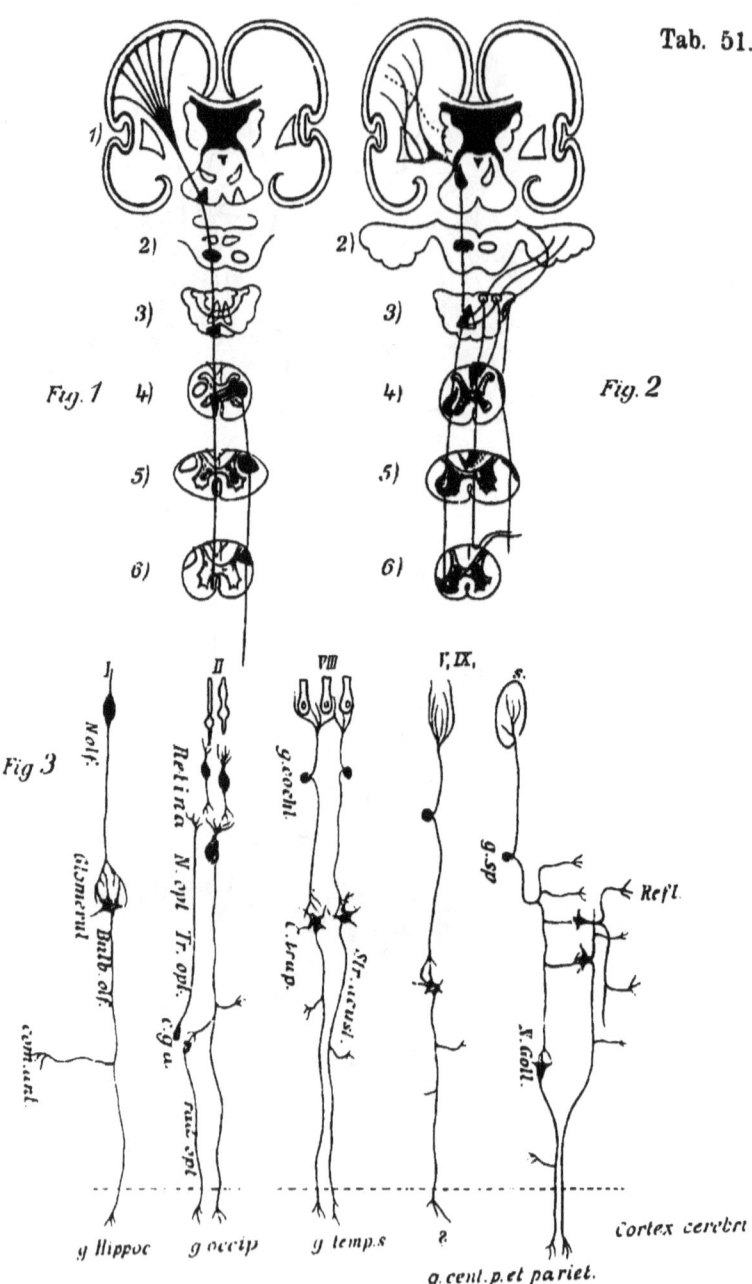

Tab. 51.

Explanation of Plate 51.

In Figs. 1 and 2 there are represented in transverse section of the hemispheres (1), the brain stem (2-4), and the spinal cord (cervical [5], lumbar cord [6]), the position of the (motor) pyramidal tract and the (sensory) fillet tract and their prolongations in the spinal cord.

FIG. 1.—*Red, the Pyramidal Tract and its Position in the Different Levels.* In 4 partial decussation: anterior pyramidal and lateral pyramidal tracts.

FIG. 2.—*Green, the Sensory Tract.*

In 6 entrance through the posterior roots. Further conduction:

a. Uncrossed in Burdach's-Goll's columns as far as their nuclei in 3;

b. Crossed (anterior commissure) and uncrossed (short tracts) after interruption in the gray substance of the antero-lateral ground bundle.

Further course: in 3, decussation of those mentioned under *a* reach the fillet after interruption in the nuclei of the posterior columns (internal arcuate fibres).

After the incorporation of the tracts mentioned under *b* (interrupted in the substantia reticularis?) the fillet passes (after receiving the central sensory tract for the cranial nerves) (for details see Fig. 3) through 2, 1 farther toward the cortex.

Brown, the (sensory) lateral cerebellar tract (see Plate 49) and fibres from the nuclei of the posterior columns in 3 to the cerebellum.

FIG. 3.—*Scheme of the Sensory Nerve Tracts.*

The peripheral sensory neuron brown.

The central sensory neuron green.

Cells, course of fibres, terminations are evident in the scheme without further explanation.

I., Olfactory tract; II., visual tract; VIII., auditory tract (cochlear nerve); V., IX., taste tract; see scheme of the cutaneous sensory tract and the related reflex tracts (shorter, longer reflex arcs).

Explanation of Plate 52.

SCHEMATIC REPRESENTATION OF THE COURSE OF THE FIBRES IN THE SPINAL CORD.

I. The Motor Tract.

a. Central neuron (red) : Lateral pyramidal tract ($Py\,l$) and anterior pyramidal tract ($Py\,a$) ; terminal arborization in the anterior horn. b. Peripheral neuron (blue) : anterior horn cells—anterior root ($r.a$)—motor nerve muscle.

II. The Sensory Tract.

a. Peripheral neuron (brown) : sensory nerve ($n.p$), spinal ganglion (Sp)—posterior root ($r.p$) of the spinal cord. In the posterior root zone of the posterior columns each fibre divides into an ascending and a descending branch (short and long fibres). The short tracts curve into the posterior horn as :
1. Reflex collaterals to the anterior horn, shorter reflex arc, longer reflex tracts (intercalation of another [green] neuron).
2. Fibres to the cells of the middle zone of the gray substance.
3. Fibres to the cells of Clarke's columns (c).
4. Fibres to the central and especially the medial anterior horn cells (commissural cells).
5. Fibres to the posterior horn cells.

The long tracts (6) pass first into Burdach's column, higher also into Goll's column, and thus to the nuclei of the posterior columns in the medulla. (Here they join the fillet.)

b. Central neuron (green). It begins with the cells of the terminal places of the peripheral, enumerated under 2 to 6.
1. From those which have been enumerated under 2 as "column cells" arise the fibres of the anterior ground bundle of the same side (fal) (fl) and the columns of Gowers (G).
2. From those mentioned under 3 : the lateral cerebellar tract of the same side (Cb).
3. From those under 4 : fibres which cross in the anterior commissure to the anterior lateral column (fal) (fl) to ascend in the other side.
4. From those under 5 : fibres to the lateral limiting layer (fl) and to the ventral field of the posterior columns.

In addition to this is represented the manner in which the collaterals are given off and the termination of the central short tracts (which quickly bend again into the gray substance) of the anterior lateral columns, the "inland cells" (green, Golgi) in the posterior horn ; the decussation in the posterior commissure is not clear. There are contained in the posterior roots apparently other individual fibres which have their neuron cells in the anterior horn, but in man this is not yet satisfactorily established.

Tab. 52.

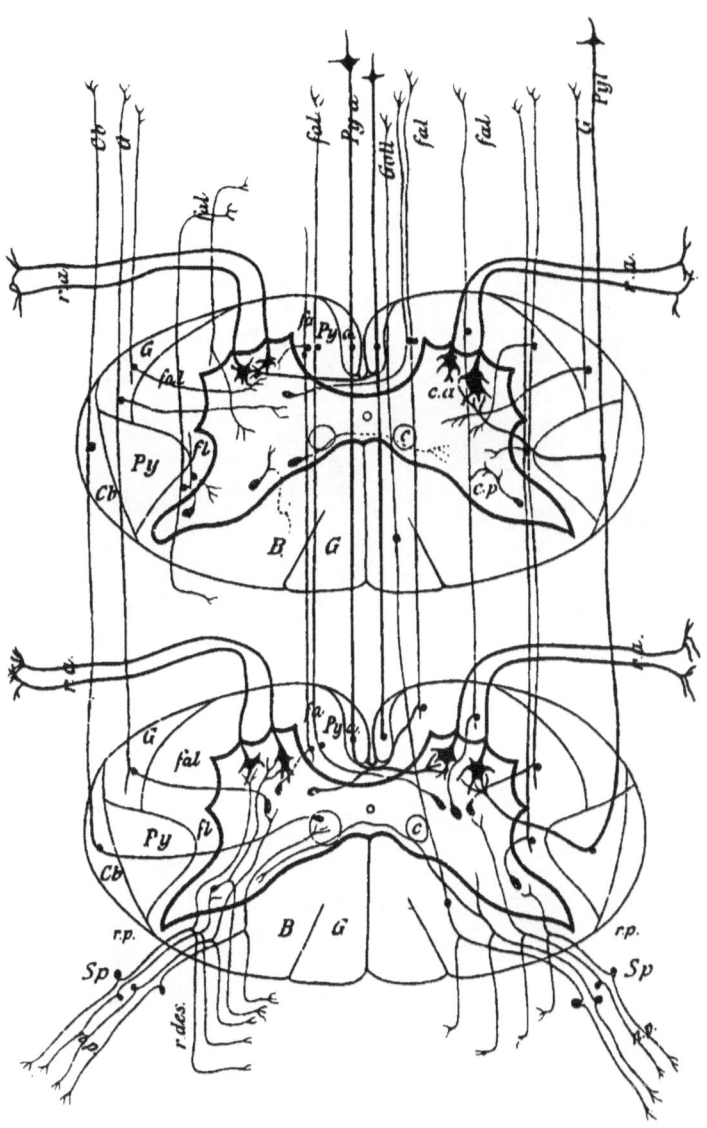

IV.

GENERAL PATHOLOGICAL ANATOMY OF THE NERVOUS SYSTEM.

Demonstrated by Special Examples.

TO ACCOMPANY SECTION IV. OF THE TEXT.

SPECIAL PATHOLOGY OF THE BRAIN.

TO WHICH BELONGS SECTION V., PAGES 150 AND THE FOLLOWING OF THE TEXT.

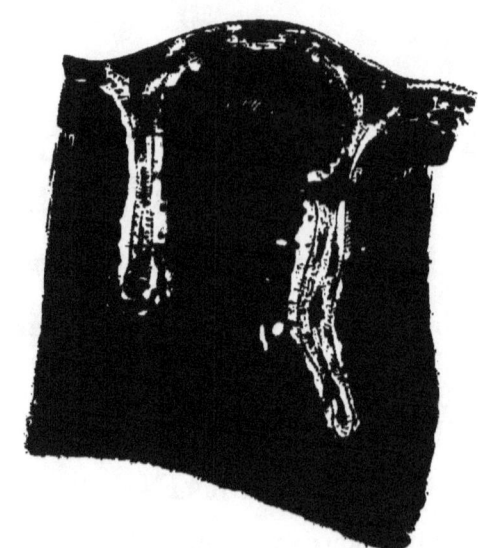

Fig 1

Fig 2

SECONDARY DISEASES OF THE NERVOUS SYSTEM.

Explanation of Plate 53.

FIG. 1.—*Section through the Cortex and Meninges of the Brain in a Case of Epidemic Cerebro-Spinal Meningitis.*

The preparation is from a soldier, 23 years old, who was taken suddenly ill with general febrile manifestations, vomiting, and headache. Stiffness of the neck, associated with pain, continuous high temperature (40° C.); herpes labialis, and leucocytosis of 36,000 established the diagnosis. Following this there was increasing disturbance of consciousness, motorial irritation symptoms such as twitchings of the face muscles and the arms, especially the right, and delirium. Death in deep coma six days later, the stiffness of the neck and trunk muscles still continuing.

Autopsy: On and in the meninges of the brain and spinal cord were found purulent, thick, white masses which covered the convexities especially. The process was most severe in the middle of the left central convolutions. In the pus the diplococcus of Fraenkel. The purulent infiltration (colored red) sunk into the fissures following the pia into the cortex. The cortical cells were very much damaged by the inflammatory products, and thus their function was inhibited.

FIG. 2.—*Cerebral Cortex from a Case of Tubercular Meningitis.*

A 30-year-old woman, who had suffered previously from tuberculosis of the lungs, was admitted twenty days before her death. Symptoms: high fever, headache, delirium, gradually developing coma. Ptosis of the eyelids, dilatation of the pupils, convulsions. Diagnosis: miliary tuberculosis.

Autopsy: the meninges studded with small whitish-gray particles. On the base of the brain an exudation in the sulci which completely embedded the motor oculi.

The preparation shows one of these tubercles in the depths of a sulcus. It has caused by its increase in size displacement of the cell layer of the cortex. A collection of such solitary tubercles may form a large tumor (especially in children). Here we have the opposite to the disseminated variety of tuberculosis.

Explanation of Plate 54.

FIG. 1.—An *aneurism* (*An*), lying closely over the corpora quadrigemina which has produced softening, especially in the region of the tegmentum of the cerebral peduncle (*T*) and the corpora quadrigemina. The patient, a man 53 years old, had the following symptoms: headache, vertigo, paresis of the left arm and leg, complete hemianæsthesia of the same side, ocular paralysis in the right eye, such as ptosis, paralysis of the sphincter pupillæ of the superior rectus, etc.

FIG. 2.—*Caries of the Spinal Column.*

The preparation from a woman, 45 years old, who became sick with pain in the back and sides, associated with increasing weakness of the legs. After three months inability to stand and the presence of mild sensory and bladder disturbances. Tendon reflexes exaggerated. Irregular fever; nothing abnormal in the spinal column. Eventually complete paralysis of the legs, contracture formations. An intercurrent cold abscess in the thorax corroborated the diagnosis of compression of the spinal cord secondary to caries of the vertebral column.

Autopsy: Carious focus in the eighth dorsal vertebra (*C*); numerous granulations and caseous masses filled the spinal canal and caused compression of the cord and the outgoing roots (*r*). *v*, body of vertebra; *pr. sp*, spinous process.

FIG. 3.—A tumor (*T*), arising from the inner surface of the dura (*d*), extending into the central canal, developed in a woman, 38 years old, with symptoms of severe pain in both legs, and in the course of three months complete spastic paralysis of both legs, as well as complete anæsthesia of the same (except cold was taken for hot) and incontinence of urine. Death resulted from cystitis and pyelitis.

Diagnosis: Compression of the spinal cord (carcinoma?). The tumor proved to be a sarcoma which completely filled the vertebral canal from the second to the sixth dorsal segments.

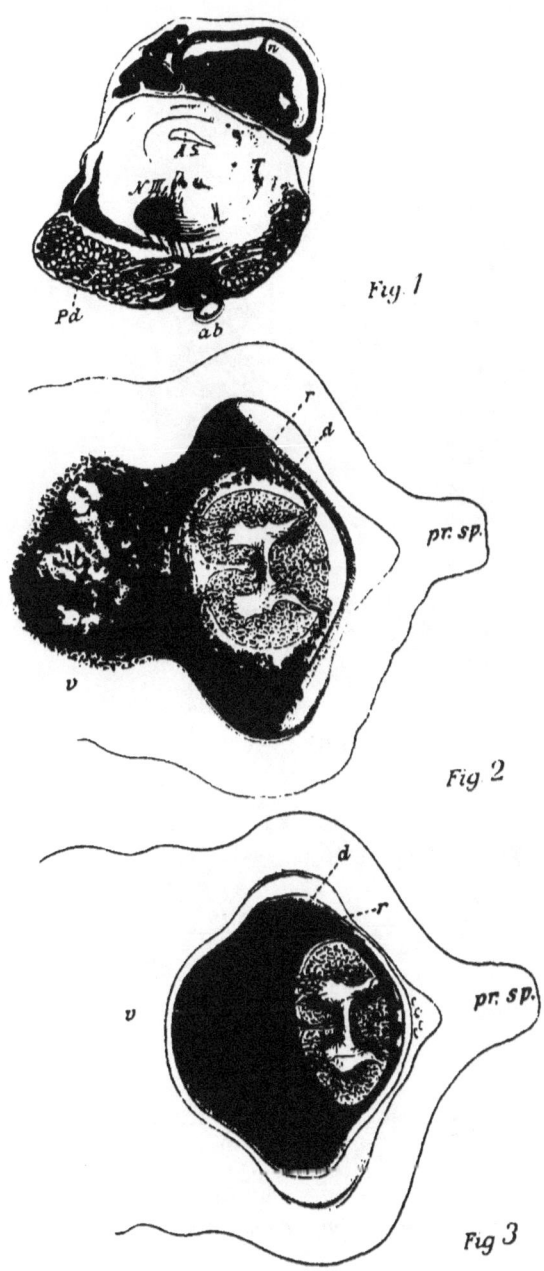

Fig. 1

Fig 2

Fig 3

Tab. 55.

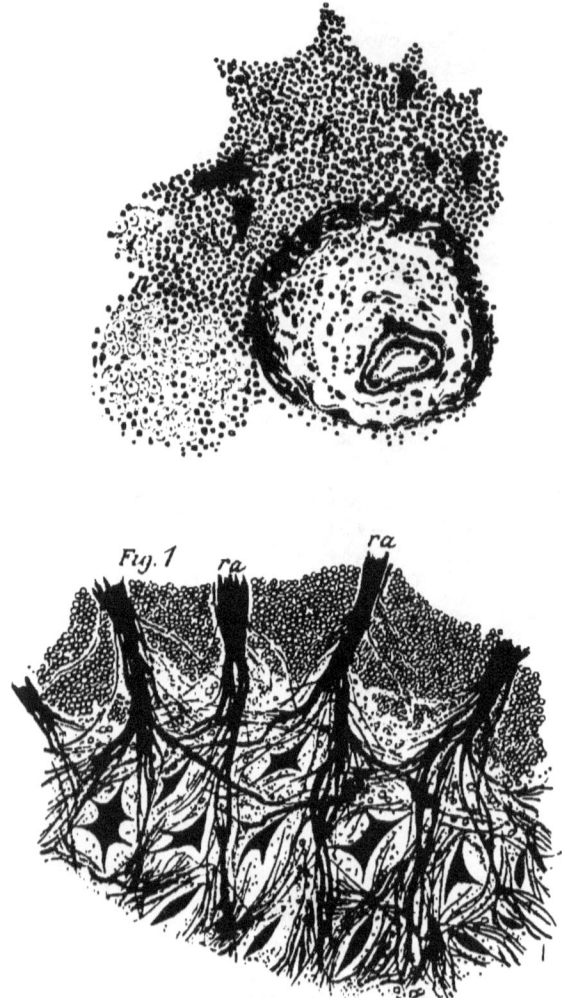

Fig. 1

ra ra ra

Fig. 2

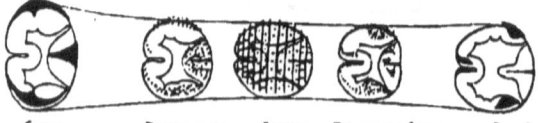

Cerv.   Degen. asc.   focus.   Degener. desc.   Lumb.

Fig. 3

Explanation of Plate 55.

Fig. 1.—*Gumma of the Base of the Brain.*

The syphiloma (*g*) has its origin from the meninges, lies in the vicinity of the adventitia of a blood-vessel (*a*), and is in part proliferated from the latter which shows the characteristic luetic endarteritis. Proliferation of the tunica intima has gone on to such an extent that the lumen of the vessels has become narrowed and nearly obliterated. The process has extended so that the neighboring branch of the vagus is encroached upon.

The preparation is from a patient, 20 years old, who had suffered repeated strokes and in whom paralysis of the left abducens and facial nerves persisted.

The autopsy showed several small gumma formations in the basal meninges, adhesive and cloudy condition of the latter, arteriosclerosis of the blood-vessels of the brain, and multiple foci of softening in the hemispheres.

PRIMARY DISEASES OF THE NERVOUS SYSTEM.

Fig. 2.—*Section through the Anterior Horn of the Cervical Cord in a Case of Spinal Muscular Atrophy.*

The cells of the anterior horn are primarily diseased, atrophied, their protoplasmic prolongations are very much less numerous than normal, and in part they have entirely disappeared (compare Plate 20, 2). The anterior roots are atrophic, secondary to the disappearance of the anterior horn cells. The degeneration extends secondarily throughout the entire extent of the neurons. There is degenerative atrophy of the motor nerves as well as of the muscles, due to the loss of the protoplasm. The nerve cells exercise a preservative influence on the fibre of the neuron and the muscles to which it is distributed.

Fig. 3.—*Scheme of Secondary Degeneration in the Spinal Cord* (see Plates 65–68).

If the continuity of a spinal-cord tract is interrupted by disease [or injury] there will result secondary column degeneration, because the neuron fibre is severed from its cell. The tracts that degenerate downward ("descending") are especially the pyramidal tracts (neuron cells in the cortex), Goll's columns (spinal ganglion cells), the lateral cerebellar tracts (cells of Clarke's columns), and Gowers' column (cells in the middle zone of the gray substance).

DISEASES OF THE GANGLION CELLS.

Explanation of Plate 56.

FIG. 1.—*Section through a Portion of the Hypoglossal Nucleus in Bulbar Paralysis.*

The nucleus is poor in fibres, its cells have become extremely small, shrunken, their protoplasmic prolongations have disappeared (compare Plate 20, 1), and in consquence of this the clinical manifestations were paralysis and atrophy of the musculature of the tongue and severe form of speech disturbance.

FIG. 2.—Cortical cells from a necrotic area of softening in the brain which was secondary to embolic occlusion of an artery; the area is seen to contain innumerable small fat bodies, products of destruction which have been stained black with osmic acid.

FIG. 3.—Nucleated cells unstained. The free fat corpuscles are carried into the protoplasm by leucocytes and in the same way transported further.

FIG. 4.—In foci which are the result of hemorrhages, coloring matter of the blood remains in the form of clumps of pigment and partly in the form of crystals stored in cells where it may be recognized a long time afterward.

FIG. 5.—Amyloid bodies are found among other places in parts of the spinal cord which are the seat of degeneration, such as in tabes. Their significance is not entirely clear.

FIG. 6.—Advanced degree of cell and fibre atrophy in the anterior horns of a spinal cord affected with amyotrophic lateral sclerosis. Many of the cells are scarcely recognizable (compare Plate 20, 2).

FIG. 7.—Acute inflammatory process in the anterior horn of the lumbar cord in a case of anterior poliomyelitis. Numerous dilated blood-vessels surround the affected ganglionic cells, which latter are undergoing partial dissolution (granular disintegration). The process terminates by the formation of scar tissue after the destruction of the nerve elements.

FIG. 8.—Scar in the anterior horn, the result of a former hemorrhage into the substance of the spinal cord (hæmatomyelia). The gray substance (cells and fibres, etc.) is destroyed and its place is occupied by cicatricial connective tissue with embedded clumps of pigment.

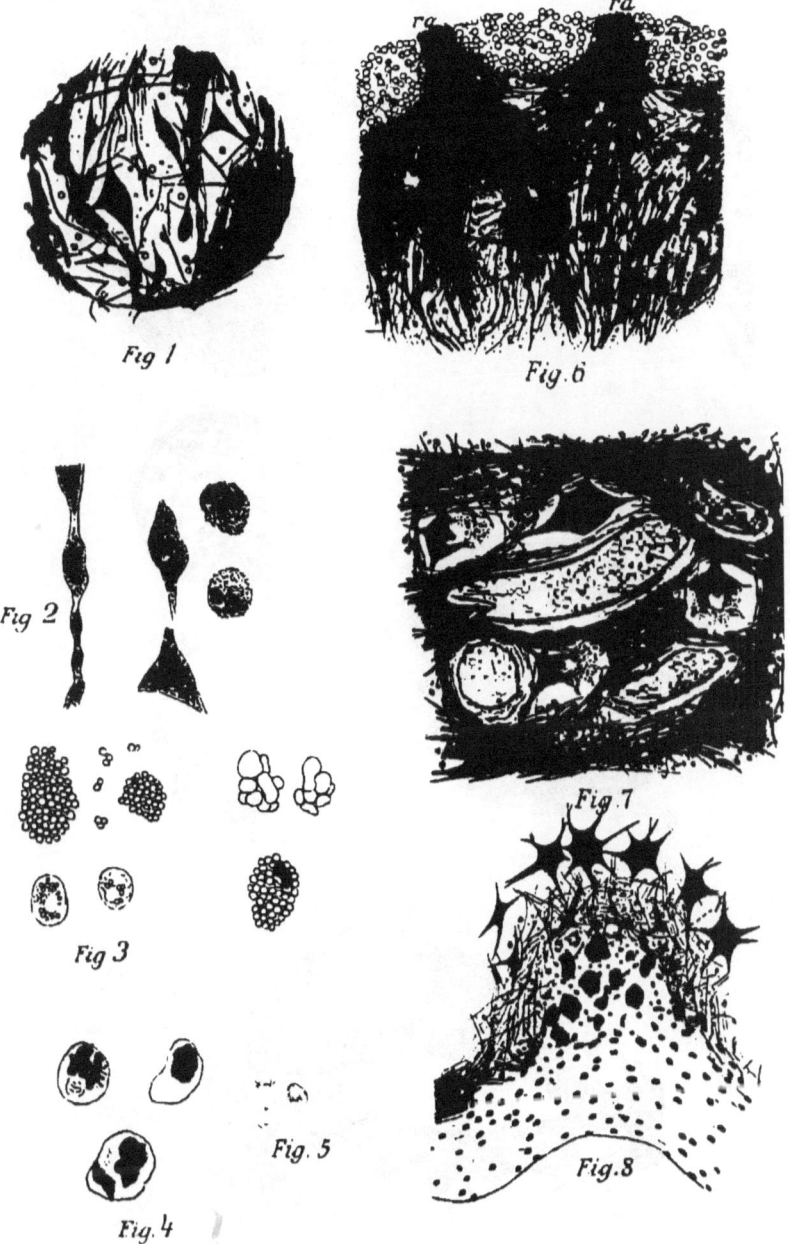

Tab 56.

Tab. 57.

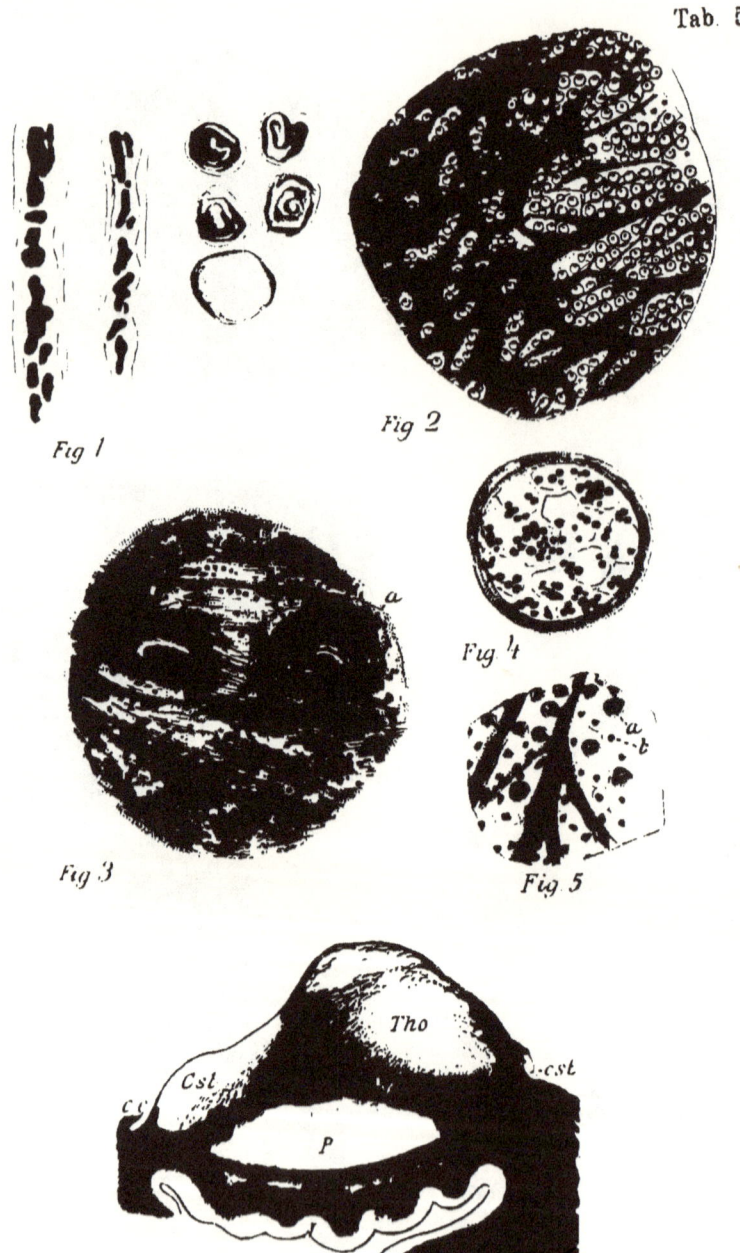

Explanation of Plate 57.

DISEASES OF THE NERVE FIBRES.

FIG. 1.—Nerve fibres whose medullary sheaths are in a degenerated condition; the fibres are swollen, the substance of the medullary sheath is disrupted, melted to drops, and in part resorbed. The axis cylinder almost always perishes first.

FIG. 2.—Section through a degenerated column of the spinal cord. The fibres which normally are closely packed together (see Plate 21, 1) are in consequence of the destruction of fibres widely separated one from another, and in the places of the degenerated nerve fibres there is seen newly formed glia tissue, the entire process being called sclerotic formation.

FIG. 3.—Myelitic scar. In a column of the spinal cord bundles of fibres have disappeared and in their place a network of glia tissue has formed which is filled with blood-vessels with thickened and infiltrated walls. This process is called total sclerosis.

FIG. 4.—Section through a small sensory cutaneous nerve in tabes dorsalis in which a portion of the medullated fibres (stained black) have perished.

FIG. 5.—In multiple sclerosis there develop foci within which the medullary sheaths of fibres are destroyed while the axis cylinder is in part preserved. Here also there occurs a formation of connective tissue in which the naked axis cylinders become embedded. *a*, medullated fibre; *b*, free axis cylinder.

FIG. 6.—Horizontal section through the internal capsule.

The cerebral motorial tracts (pyramidal tracts) have here suffered primary degeneration. Weak magnification of these preparations made with a medullary sheath stain shows the degeneration as light, clear areas in the middle of the surrounding normal black-stained fibres. These areas are also the places in which these tracts traverse the internal capsule and which are here clearly marked out. (Compare Plate 30, middle.)

Explanation of Plate 58.

DISEASES OF THE MUSCLE FIBRES.

FIG. 1.—Normal transverse section of muscle in combination with transverse section of polygonal muscular fibre of small size with few cells.

FIGS. 2 and 3.—Transverse and longitudinal section of muscles from a case of spinal muscular atrophy (biceps muscle). There is great dissimilarity in the thickness of individual fibres. One part has entirely disappeared, another is very much smaller (p). The muscle nuclei are very much increased and constitute rows of nuclei. The transverse striation of the protoplasm has disappeared in the degenerated fibrillæ and the protoplasm has become granular.

FIG. 4.—Juvenile muscular atrophy (progressive muscular dystrophy. Quadriceps muscle). Here whole bundles of muscle fibre have disappeared, the nuclear proliferation is striking, and many fibres are hypertrophic. Transverse striation remains for a long time in spite of the extensive atrophy of the fibres. Frequently there is a deposition of fat between the muscle fibrillæ that are preserved (pseudo-hypertrophy). At h a so-called hypertrophic muscle fibre.

FIG. 5.—Neurotic form of muscular atrophy (opponens pollicis muscle). One recognizes the great lessening in size of the fibrillæ and the nuclear proliferation. The transverse striations remain clear for a long time.

FIG. 6.—Unstained muscle fibre:

a. Degenerative muscle atrophy from the quadriceps muscle in a case of lumbar myelitis. There is a granular, cloudy condition of fibrillæ, the transverse striations have disappeared, the fibres have become irregular and small. Toward the end there is a resorption of the destroyed protoplasm, the empty sarcolemma sheath remaining.

c. Slight neural muscle atrophy, small fibres with nucleated cells and preserved transverse striation.

Tab. 58.

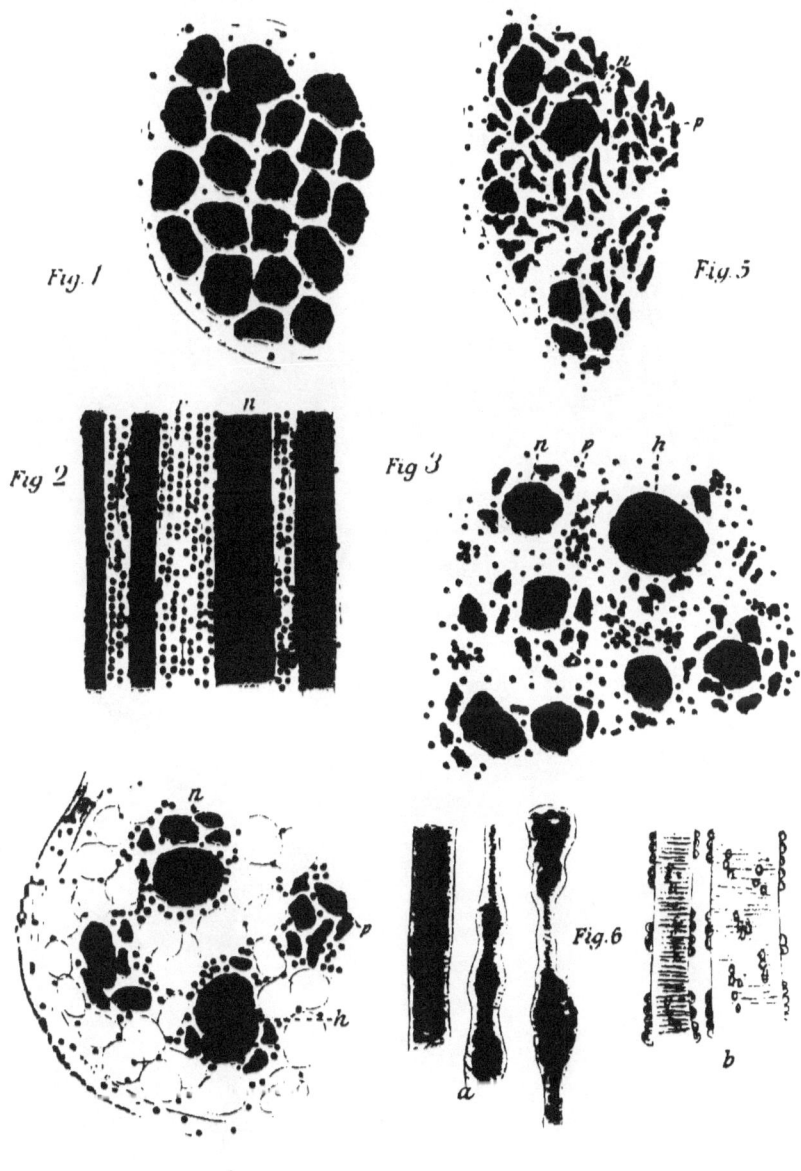

Tafel 59.

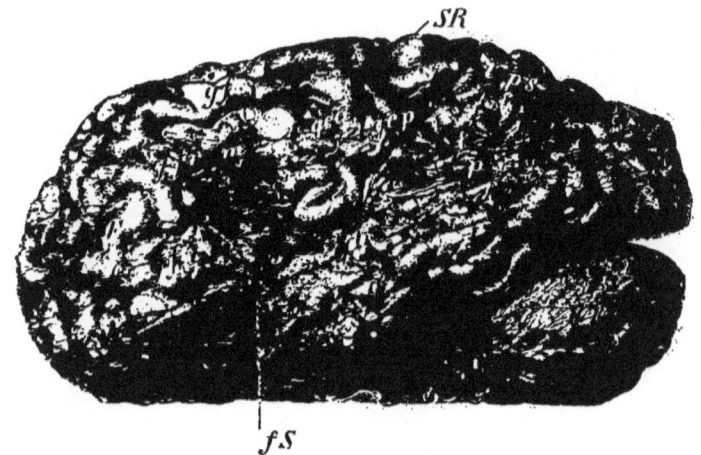

Fig. 1.

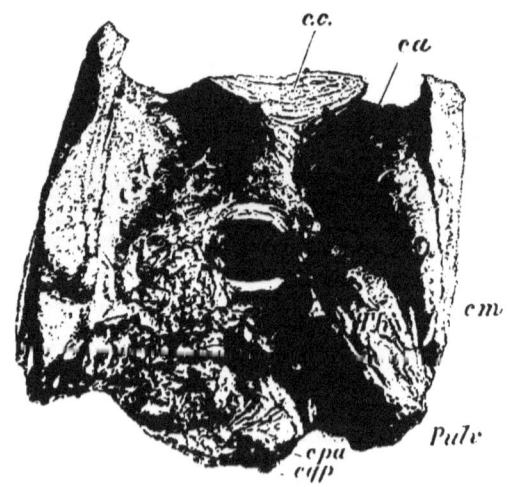

Fig. 2.

Explanation of Plate 59.

Fig 1.—*The left hemisphere* of a man, 56 years old, who developed after repeated strokes a spastic paralysis of the right facial and hypoglossal nerves, of the right arm, and a weakness of the right leg (right-sided hemiplegia). In addition to this there was a loss of voluntary speech (motor aphasia), defective understanding of words and of writing (word dumbness, loss of optic memory pictures). After three years, during which time dementia increased, he died.

The photograph shows the outer surface of the left hemisphere, a great part of the cortical convolutions is destroyed (porencephaly). Entirely transformed into connective tissue are: the posterior portion of the middle and the inferior frontal convolutions (*g.fr.m* and *i*), the upper and middle temporal convolutions (*g.t.s* and *m*), the inferior parietal lobe (*l.p.i*), and in great part the anterior and posterior central convolutions, especially in their inferior portions (*g.c.a* and *p*), and the middle of the occipital lobes (*o*). Under the cortex the island of Reil and the medullary substance are extensively destroyed. Fossa of Sylvius (*fS*), fissure of Rolando (*SR*).

Fig. 2.—View of the third ventricle opened (the corpus callosum and fornix removed). Complete destruction of the left thalamus and its environs (corpora quadrigemina, internal capsule).

A man, 50 years old, had a stroke; three days later when he regained consciousness there was right-sided hemiplegia (face-arm-leg). Simultaneously there occurred a paralysis of the motor oculi of the left eye (hemiplegia alternans). Speech was not disturbed, but there was right-sided hemianæsthesia.

The extensive focus was the result of a hemorrhage. Beneath the thalamus opticus of the left side the subthalamic region and outside it the posterior limb of the internal capsule were destroyed. In addition there was destruction of the left corpora quadrigemina (*cqa*) and the region of the tegmentum with the left motor oculi nucleus. (*Th*, thalamus; *Pulv*, pulvinar; *cm*, median commissure in the middle; *ca*, anterior horn of lateral ventricle; *Cst*, corpus striatum; *c.c.*, corpus callosum).

Explanation of Plate 60.

FIG. 1.—*Section through the anterior quadrigeminal bodies* in complete destruction of the left tegmental region (*t*); the left quadrigeminal body is atrophic. The nuclear region of the left motor oculi beneath the aqueduct of Sylvius (*AS*) is destroyed. The tegmental region, the red nucleus, superior and inferior fillet, substantia recticularis, etc., are entirely absent and the left crusta is completely atrophic. Clinical condition as in case of Plate 59, Fig. 2; alternating hemiplegia.

FIG. 2.—*Section through the medulla* of a young girl who had inherited a tuberculous tendency. She became sick rather suddenly with headache, vomiting, increasing weakness of the right leg and later of the right arm. Soon after there occurred paralysis of the facial and of the hypoglossal nerves on the left side (hemiplegia alternans), paralysis of swallowing, and sudden death.

The autopsy showed a tumor of the pons, which had extended far into the medulla (tubercular) and had thus destroyed the pyramids and outgoing facial and hypoglossal nerves. The tumor was a solitary tubercle with an enormous number of giant cells.

FIG. 3.—We have considered previously *focal diseases* of the brain. The accompanying preparation shows a system disease of the same: *chronic progressive ophthalmoplegia*. The cells of the motor oculi nucleus (*NIII*) are primarily degenerated and the normal number of fibres of the outgoing motor oculi nerve is greatly diminished (*III*) (compare normal section Plate 35, 2).

The disease occurred in a woman sick with tabes. It caused a slowly developing paralysis of all the eye muscles.

(*c.p*, posterior commissure; *Sn*, substantia nigra; *nr*, red nucleus.)

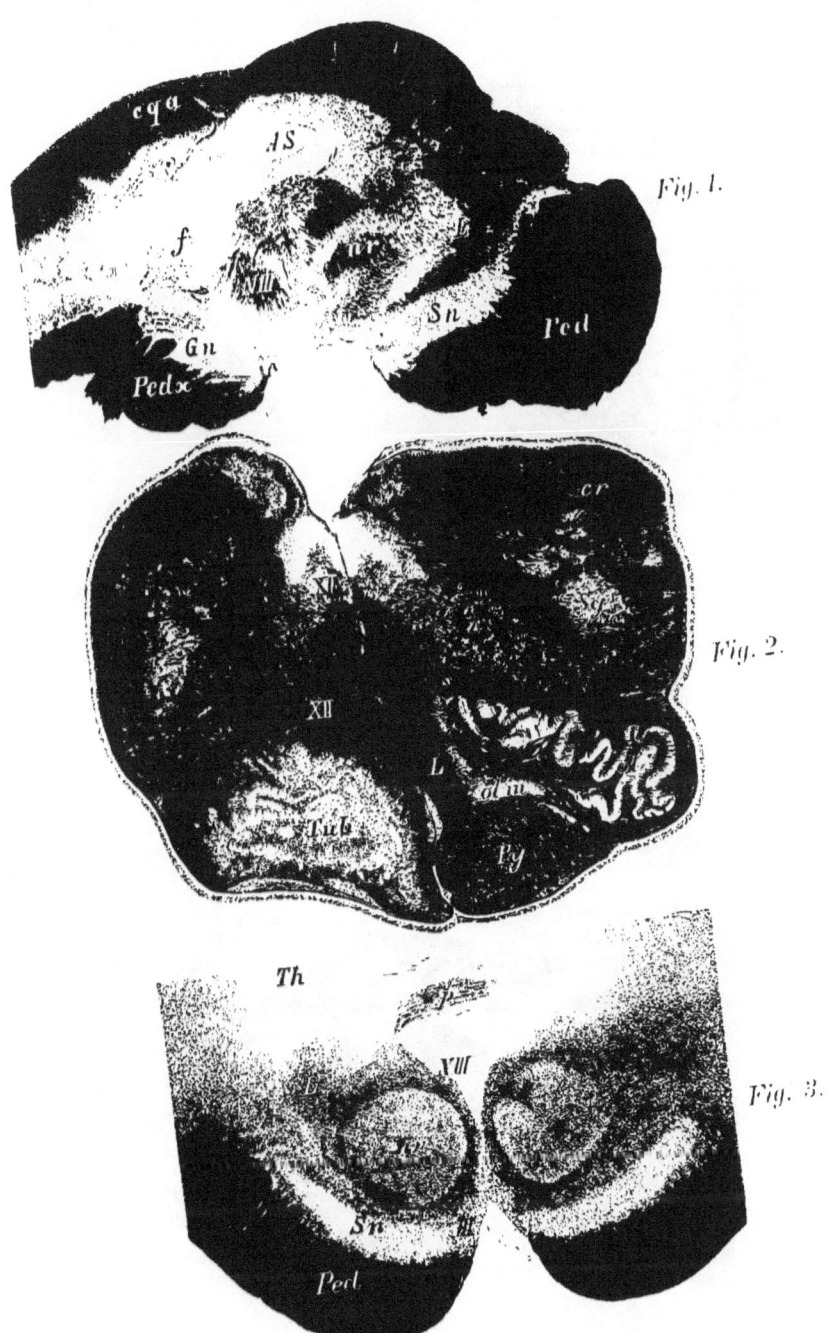

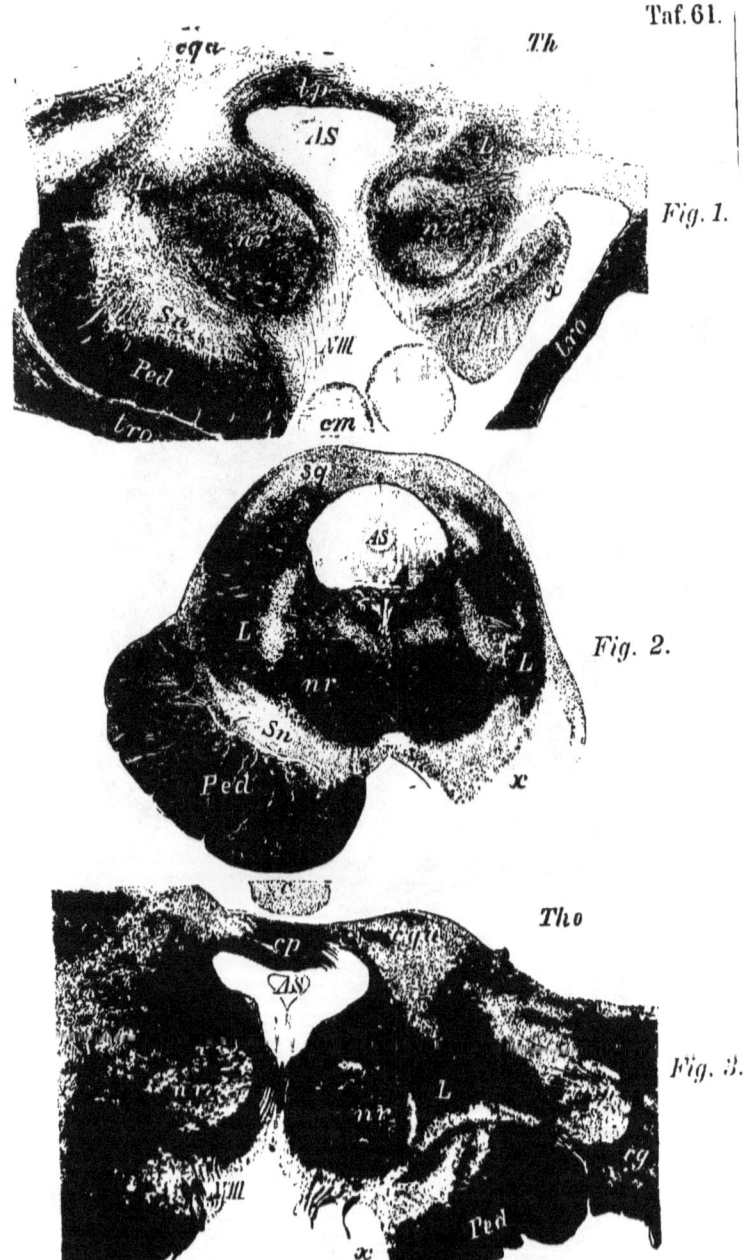

Taf. 61.

Fig. 1.

Fig. 2.

Fig. 3.

Explanation of Plate 61.

SECONDARY DEGENERATION IN THE CRUSTA.

FIGS. 1 and 2 are cross sections through and closely beneath the anterior corpora quadrigemina. There is a complete atrophy of the right pes pedunculus cerebri (x). The preparation is from a case of extensive primary destruction of the cortex of the brain. As the fibres of the peduncle have their neuron cells in the cortex, it follows that after the degeneration of the latter a secondary descending degeneration must follow.

The fillet tract is smaller on the right side than on the left, but it is not totally degenerated (simple, not degenerative atrophy). The fillet fibres have their neuron cells not in the cortex, but partly in the nuclei of the posterior columns and partly in the subthalamic region and in the thalamus. Clinically there had been total left-side hemiplegia).

(tro, optic tract; Sn, substantia nigra; L, lemniscus; cqa, anterior corpora quadrigemina.)

FIG. 3.—Section through the anterior quadrigeminal bodies in total secondary atrophy of the most mesial peduncle bundles (frontal pontine tract) (x). This partial pes pedunculi degeneration resulted secondarily to an area of embolic softening which had destroyed the limb of the capsule and the lenticular nucleus.

Clinically there had been no indicative symptoms.

(cgm, medial geniculate body; cg, lateral geniculate body; r, pineal gland, etc.)

These and the following examples of secondary degeneration in the brain and spinal cord are for neurology especially of great significance, because it is by the course of such degeneration that the different tracts have been made certain. We utilize the experiments made on animals inferentially in man. On such grounds have been developed the considerations here given. It forms a good repetition of what is given in Section III.

Explanation of Plate 62.

Secondary Degeneration in the Cerebral Peduncles.

Fig. 1.—*Section through the Third Ventricle* (middle commissure). Beneath the lower border of the thalamus which has been severed above (*Th*) is situated the remaining portion of the subthalamic region (Luys' body) (*cL*), medullary substance of the red nucleus (*L*), beneath it the internal capsule (posterior limb) before its exit as pes pedunculus cerebri. In the middle part of the latter (*x*) is an area of degeneration secondary to a destructive focus (hemorrhage) in the central convolutions. The degenerated area represents a part of the pyramidal tract.

Fig. 2.—*Section through the Anterior Corpora Quadrigemina of the same Case.* One recognizes here the position of the degenerated pyramidal tract (*Py*) in the middle of the pes pedunculus cerebri; it involves the entire pathway (not totally degenerated here) and in addition a large area lateral to the degenerated portion.

Clinically there had been left-sided hemiparesis of the face, arm, and leg.

(*tro*, optic tract; *f*, the descending pillar of the fornix which passes backward to the mammillary body; *AS*, aqueduct; *cL*, subthalamic body, etc.)

Tafel 62.

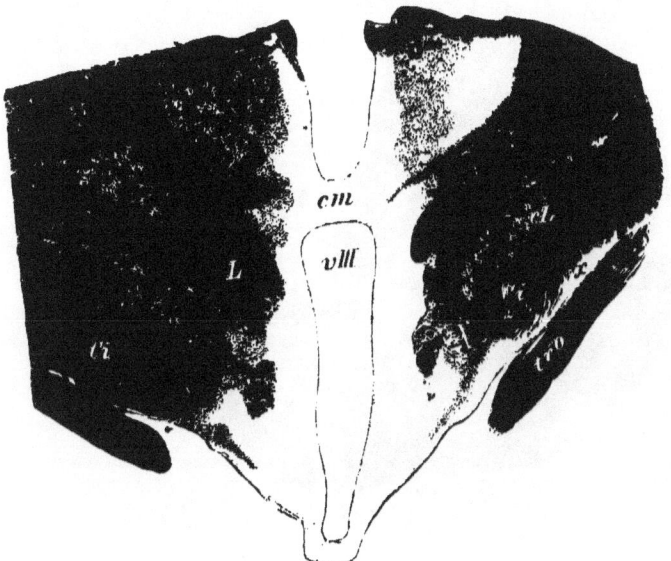

Fig. 1.

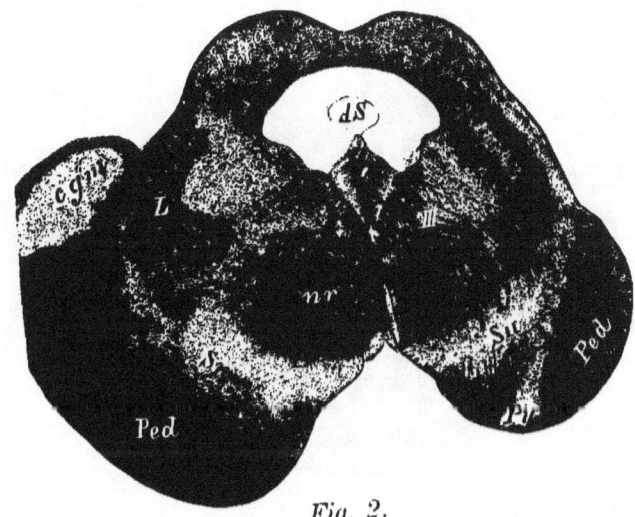

Fig. 2.

Tafel 63.

Fig. 1.

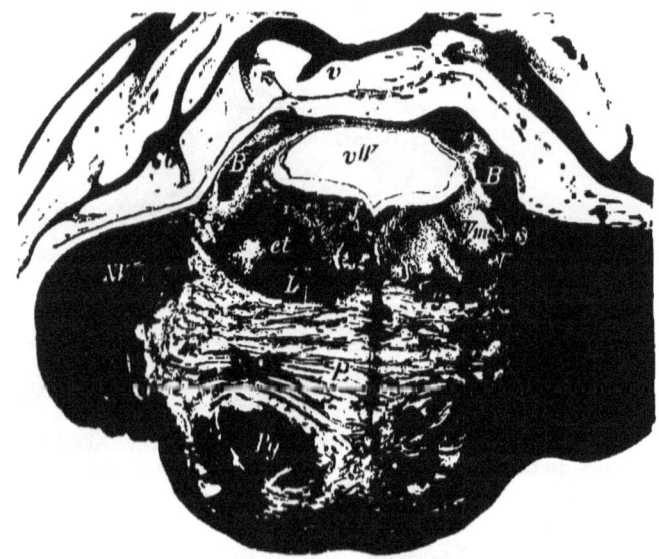

Fig. 2.

Explanation of Plate 63.

SECONDARY DEGENERATION IN THE PONS.

FIG. 1.—*Section through the Anterior Half of the Pons.* The section belongs to the case Plate 61, 1 and 2. The fibres of the crusta on the left pass through in many bundles between the transverse fibres of the pons; on the right these are absent, as are also the ganglia of the pons, and to this is due the enormous shrinkage of the right half of the pons. The tegmental region is normal. It is, however, to be seen that the secondary degeneration does not confine itself to the fibres that lie between two directly connected neuron ganglion systems, but involves their cells as well.

FIG. 2.—*Section through the middle of the pons*, with extensive secondary degeneration in the tegmental region. The preparation is from the case Plate 59, 2 (left-right!). Both brachia conjunctiva (B) (in part, right at z) (neuron cells in the destroyed red nucleus), are degenerated; then in the right tegmentum the central tegmental tract (ct), the entire upper fillet (y) (neuron cells in the destroyed subthalamic region and in the thalamus?), the nasal trigeminus root (Vn), the right pyramidal tract (x) in part, and other tracts are degenerated.

(Vm, motorial; Vs, sensory trigeminus nucleus and between the two the descending root (Vc); f, posterior longitudinal fasciculus; Ls, superior; Li, inferior fillet.)

Explanation of Plate 64.

SECONDARY DEGENERATIONS IN THE MEDULLA.

FIG. 1.—*Section through the Medulla just behind the Pons.*

The left pyramidal tract (x) shows total secondary degeneration, the result of primary destruction of the posterior limb of the internal capsule by a hemorrhage (right-sided hemiplegia).

One sees here the well-known course of the left striæ acusticæ passing from the ventral acoustic nucleus decussating over the floor of the fourth ventricle (to the fillet).

FIG. 2.—*The Medulla from a Case of Infantile Cerebral Palsy.*

The preparation is from a man, 30 years old, who developed in his early youth (congenital? acquired?) a cessation of development of the entire right side of the body. There was flexion contracture, and paresis of the right arm and weakness of the right leg. Epileptiform attacks, athetosis.

		Upper arm.	Forearm.
Length of arms,	Right, 29 cm. Circumference, Left, 31 "	Right, 21 cm. Left, 24 "	Right, 19 cm. Left, 24 "

Circumference of thigh,	Right, 35 cm. Left, 39 "	Calf,	Right, 29 cm. Left, 32 "

Autopsy: The left central convolutions are shrunken (agenesis?), the left pyramid (x) entirely atrophied, likewise the left fillet without strict degenerative loss of fibres, due to the fact apparently that the disease confined itself to one side, as the medullary sheaths were also involved. The raphe was entirely displaced toward the left side and the right pyramid almost hypertrophied (compensation?).

FIG. 3.—*Medulla Oblongata, Case of Atrophy of the Fillet.*

The right fillet (Lx) is degenerated in a descending direction and with it the internal arcuate fibres (fai, y) of the substantia reticularis (Sr) of the left side! These arciform fibres form the direct continuation of the crossed fillet and connect it with the nuclei of the posterior columns (nG) which pass up in the restiform bodies (Cr). Pyramids ($Py\ x$).

The clinical condition was hemianæsthesia of the left side of the body.

Tafel 64.

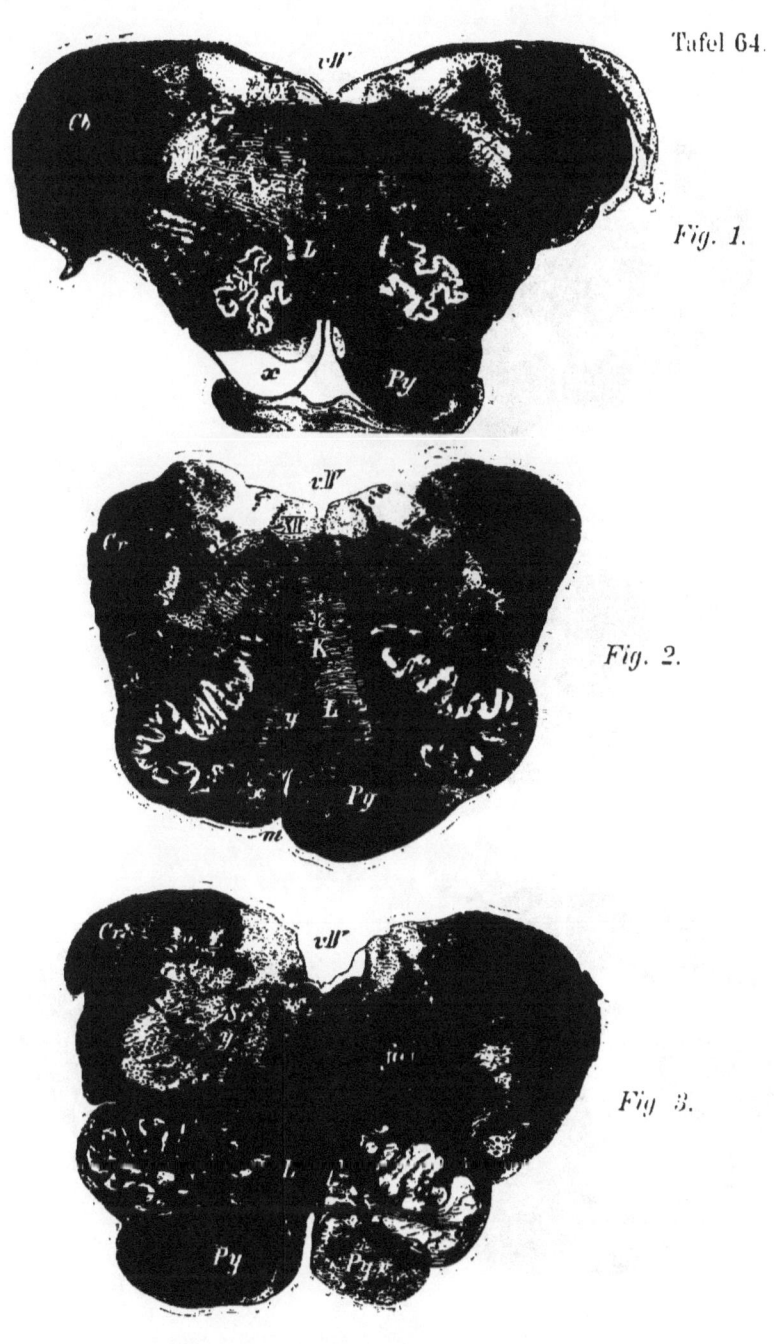

Fig. 1.

Fig. 2.

Fig. 3.

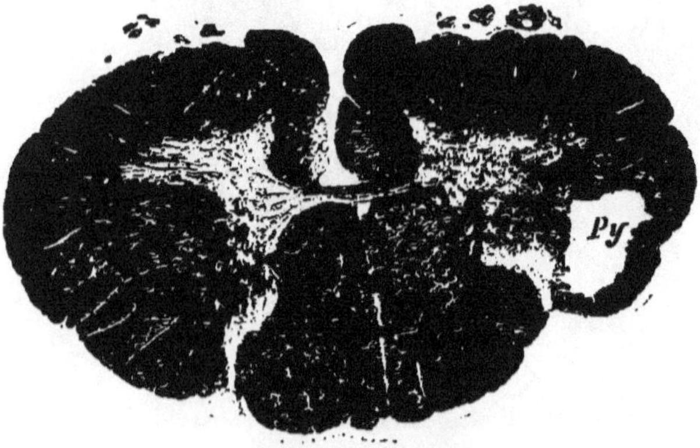

Taf. 65.

Fig. 1.

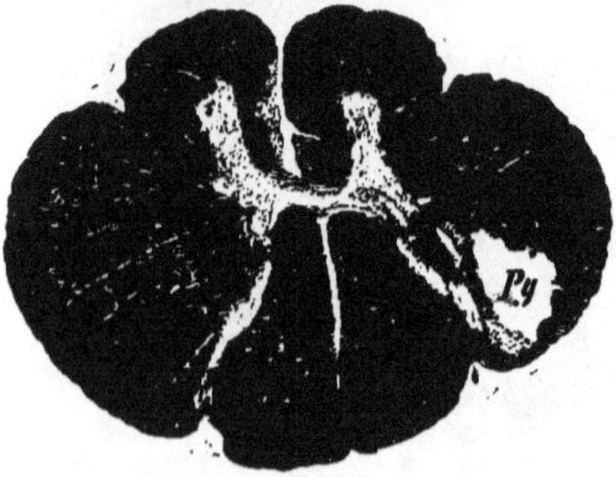

Fig. 2.

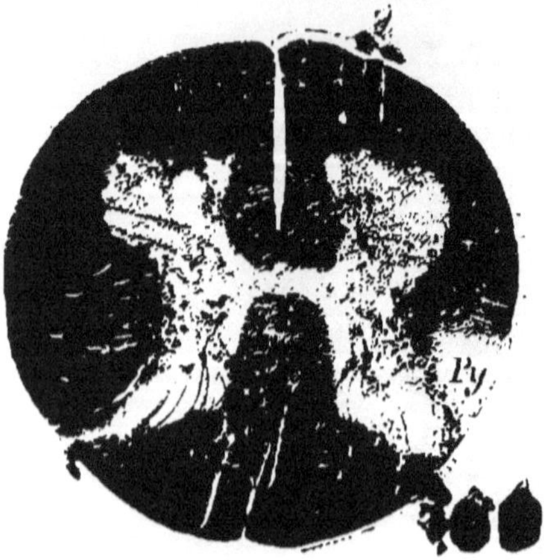

Fig. 3.

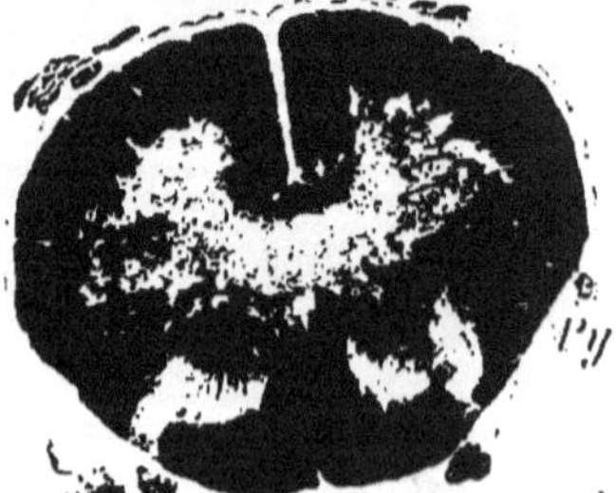

Fig. 4.

Explanation of Plate 65.

DESCENDING DEGENERATION OF THE PYRAMIDAL TRACT IN THE CORD FROM A FOCUS IN THE BRAIN.

The preparations belong to the case Plate 64, Fig. 1. The left pyramid is totally degenerated (secondary), and this degeneration may be followed in this tract throughout its entire extent in the spinal cord, also through the pyramidal crossing of the lateral tract (crossed tract) and in the anterior tract of the same side (uncrossed tract).

The lateral pyramidal tract (Py) is sharply defined from the normal lateral cerebellar tract in the cervical and dorsal region, but in the lumbar cord the pyramidal tract approaches the periphery, giving a wedge-shaped appearance. The anterior pyramidal column (x) is degenerated as far as the lower dorsal segments.

In the anterior horn no change of any significance is to be seen. The degeneration is confined to the central motor neuron, the pyramidal tract, in its entire course.

FIG. 1.—Cervical cord (level sixth cervical nerve).
FIG. 2.—Dorsal cord (level third dorsal nerve).
FIG. 3.—Lumbar cord (level second lumbar nerve).
FIG. 4.—Sacral cord.

Explanation of Plate 66.

DESCENDING DEGENERATION IN THE SPINAL CORD IN SPINAL-CORD DISEASES.

FIG. 1.—*Lower Cervical Cord.*
FIG. 2.—*Lower Dorsal Cord.*
FIG. 3.—*Upper Lumbar Cord.*

The preparations are from a man, 40 years old, who suffered a fracture of the cervical spinal column, the result of a fall, which caused crushing of the lower cervical cord. There resulted: complete spastic paraplegia of the legs, muscular atrophy and paralysis of the small muscles of the hands, anæsthesia, incontinence of urine, etc.

Autopsy: The lower half of the area of compression involved the cervical cord: descending degeneration of both lateral pyramidal and anterior pyramidal tracts, as well as a comma-shaped area (Schultze's comma, ascending fibres arising from the gray substance) in Burdach's columns, also a pale area in Goll's columns. It is to be observed that to the left lateral pyramidal column tract there belongs a small anterior column tract, to the right a not sharply defined anterior column tract (individual peculiarities). In the dorso-lumbar cord there is double-sided degeneration of the pyramidal tracts. The degeneration of the posterior columns has disappeared (short tracts).

FIG. 4.—*Lumbar Cord in Compression of the Dorsal Cord.*

The preparation is taken from the same case as Plate 54, Fig. 3, a sarcoma of the dura. Below the point of compression there occurred a degeneration in the pyramidal tracts of both sides and a slight degeneration in the border zone of the anterior lateral tracts (descending branches from the gray substance [middle zone] of the central sensory neurons).

The large number of fibres radiating into the anterior horns from the posterior horns and anterior roots is clearly represented.

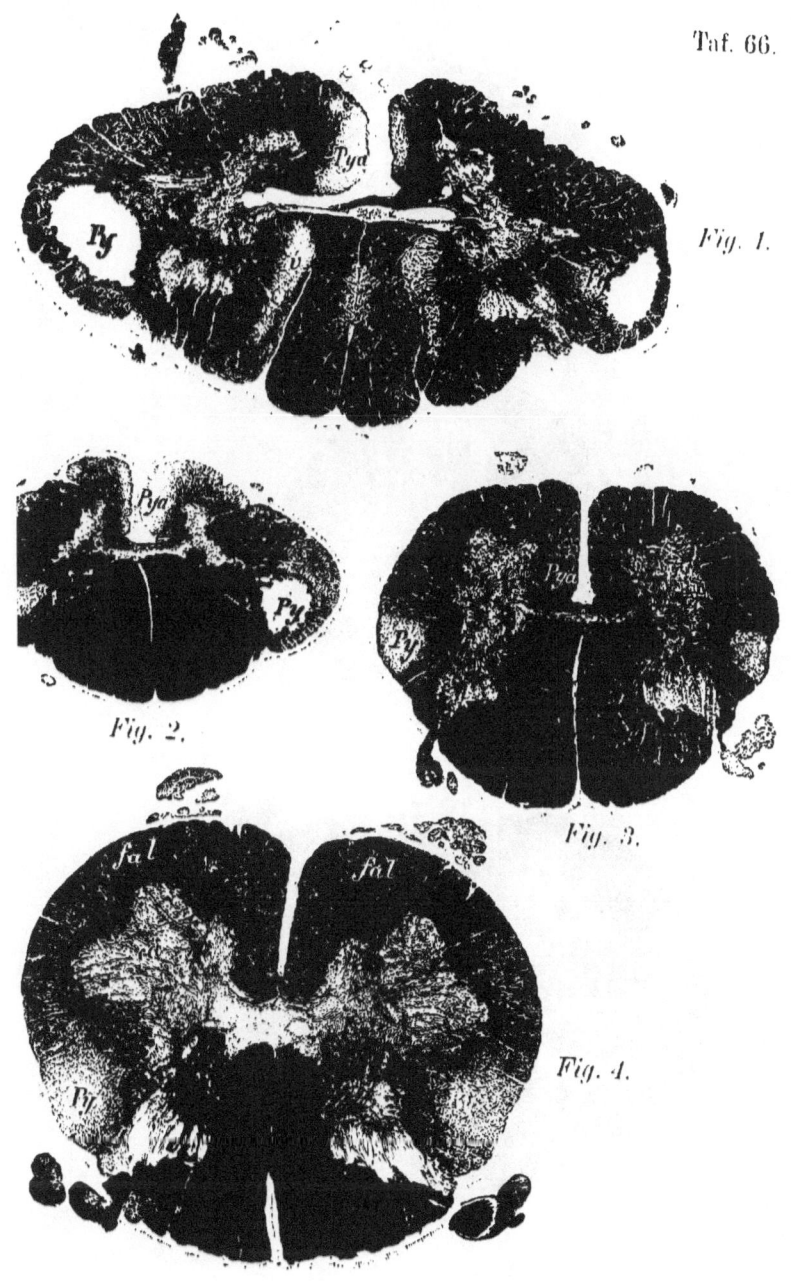

Taf. 66.

Fig. 1.
Fig. 2.
Fig. 3.
Fig. 4.

Tafel 67.

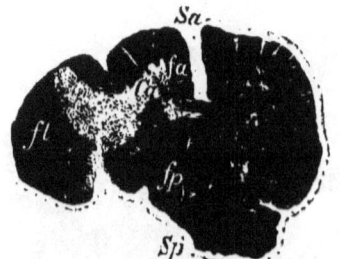
Fig. 1.

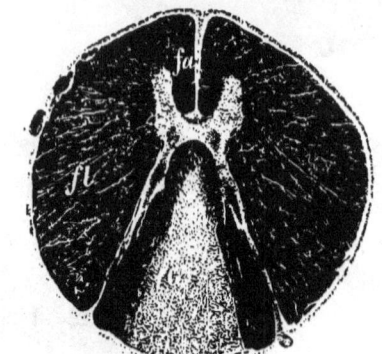
Fig. 2.

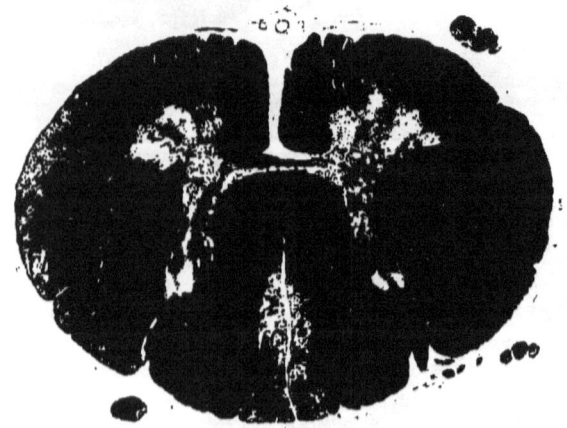
Fig. 3.

Explanation of Plate 67.

ASCENDING DEGENERATION IN THE SPINAL CORD.

A soldier, 22 years old, received a severe fall on the lower dorsal spinal column which resulted in fracture of the bone and transverse crushing of the lumbar cord. The consequences were a complete flaccid paralysis of the lower extremities, which was followed by rapidly developing muscular atrophy of a degenerative type and disturbances of sensibility, paralysis of the sphincter of the bladder and rectum, and loss of patellar reflex. Death from cystitis two months later.

FIG. 1.—*The Crushed Upper Lumbar Cord.*

FIG. 2.—*Middle Dorsal Cord.* Here is to be seen a secondary ascending degeneration of the columns of Goll (neuron cells in the spinal ganglia).

FIG. 3.—*Lower Cervical Cord.* Goll's columns (*fG*) show a wedge-shaped area of degeneration; their surface is very much smaller than in sections from a lower level.

In addition there is to be seen here an ascending degeneration of the left lateral cerebellar tract (*Cb*) and of the left column of Gowers (*G*). (As can be perceived in Fig. 1, the gray substance and Clarke's columns are very much more extensively destroyed than the right; here lie the neuron cells for those columns.)

Goll's columns conduct accordingly the long sensory tracts from the lower extremities upward; these fibres lie laterally to those that enter higher up, the long tracts from the trunk and upper extremities (Burdach's columns).

Explanation of Plate 68.

ASCENDING DEGENERATION IN THE CERVICAL CORD
AND THE MEDULLA.

FIG. 1.—*Ascending Secondary Degeneration in the Cervical Cord* consequent to primary acute dorsal myelitis. As the result of the interruption of continuity of the fibres in the dorsal cord, the following columns degenerate in an ascending direction on both sides: Goll's columns (fG), the lateral cerebellar columns (Cb), and Gowers' columns with their posterior gussets.

FIG. 2.—*Section through the Pyramidal Crossing.*

Ascending degeneration in chronic (syphilitic?) myelitis of the dorsal cord. Goll's columns (fG), the lateral cerebellar columns (Cb), and Gowers' columns (G) are degenerated.

FIG. 3.—*Section through the Medulla* at the level of the inferior olives.

The preparation is from the same specimen as Plate 67 and follows immediately after the section of Fig. 3.

One can recognize the ascending degeneration of Goll's columns (fG) which at this level have passed into the nuclei of Goll (nG). The degeneration has not affected the terminal fillet fibres (a new individual neuron!). The degeneration of the left lateral cerebellar tract is clearly apparent (Cb).

In consequence of the somewhat oblique course of the section only the left olive is included.

(L, fillet; fai, internal arcuate fibres; olm, median olive, etc.)

Tafel 68.

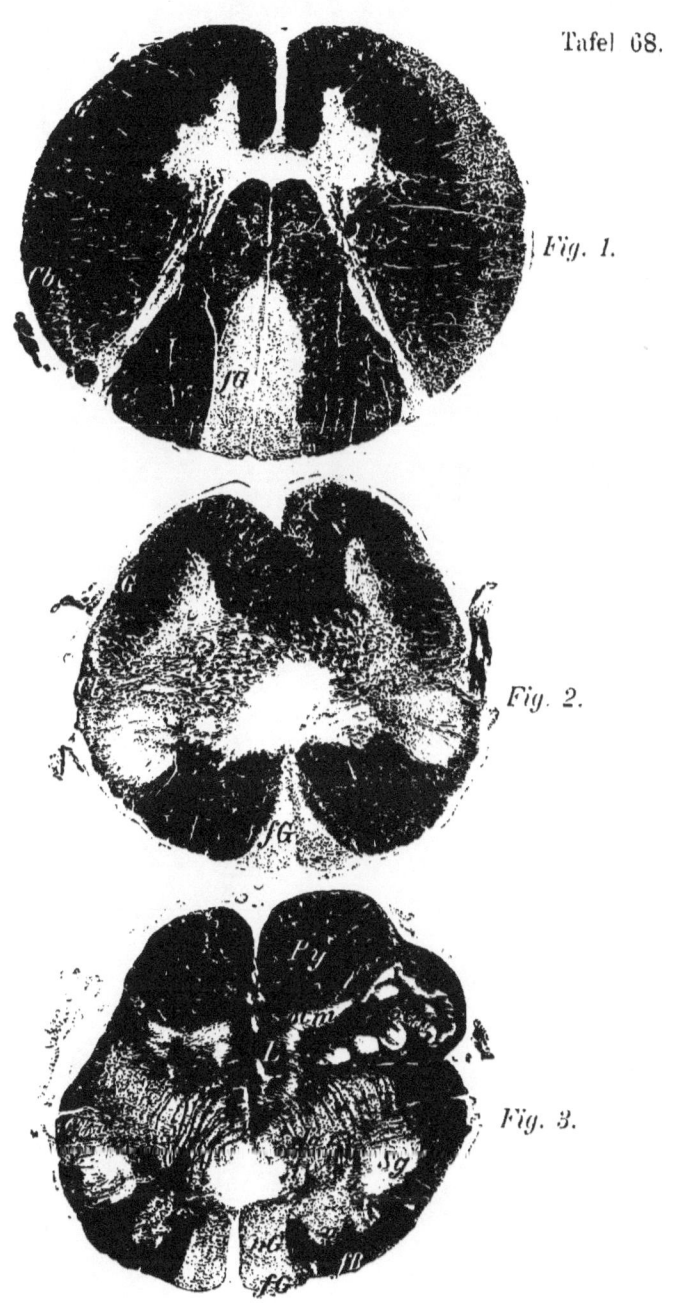

Fig. 1.

Fig. 2.

Fig. 3.

V.
SPECIAL PATHOLOGY OF THE SPINAL CORD AND OF THE PERIPHERAL NERVES.

(To which Belongs Section V., Pages 188 et seq. of the Text.)

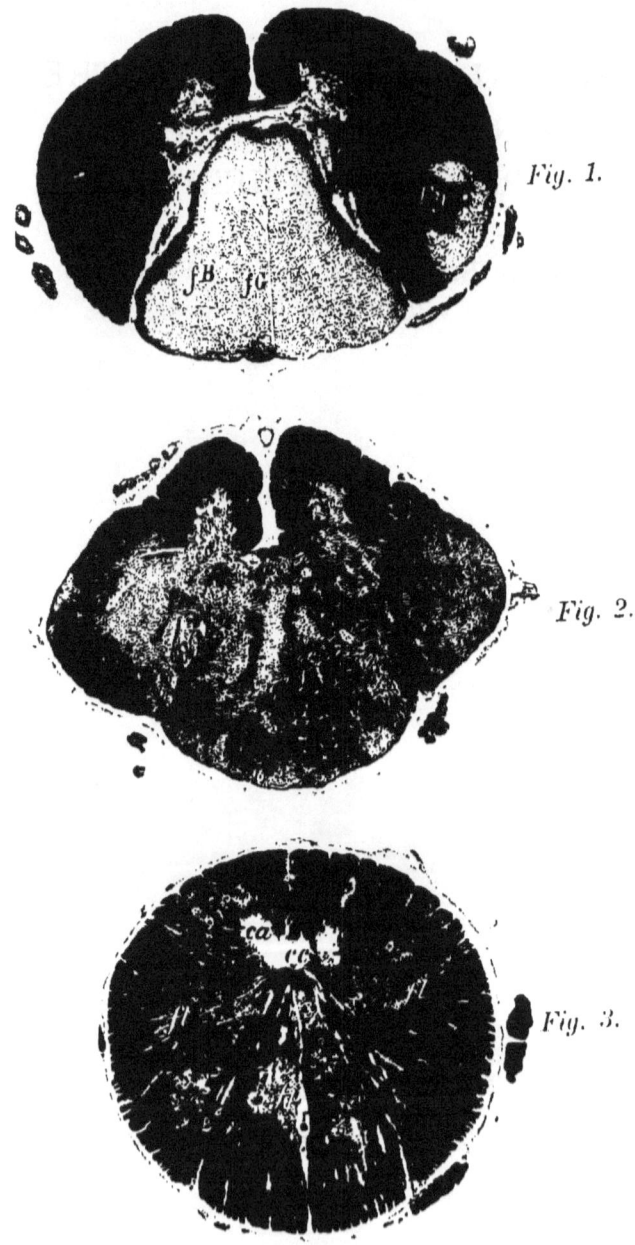

Explanation of Plate 69.

Forms of Myelitis.

Fig. 1.—*Acute Column-Form Myelitis (Toxic). Cervical Cord.*
A man, 42 years old, developed sequentially to a severe erysipelas, first painful sensitiveness of the arms and legs, which was quickly followed by paresis but not complete paralysis, slight disturbance of sensibility and of the function of the bladder (pain sensation normal), terminal diaphragm paresis. Duration of disease only four months.
Diagnosis: Multiple neuritis.
Autopsy: Extensive myelitic degeneration of the posterior columns (fG, fB), which reached from the cervical to the lumbar cord region; myelitic foci in the lateral columns, especially of the cervical cord; slight changes in the nerves.
The case would seem to indicate that conduction through the posterior columns for the sensibility of the skin is not necessary.

Fig. 2.—*Chronic Myelitis (Syphilitic?) of the Dorsal Cord.*
A 48-year-old man (probably syphilitic) sickened with slow, progressive weakness of the legs, no pain. After eighteen months complete paraplegia, sensibility in all qualities lost; patellar reflexes lively; incontinence.
Autopsy showed a diffuse, widespread myelitis extending over the entire transverse section of the cord, its severest manifestations being between the seventh and ninth dorsal segments. Beneath this the pyramidal tracts were the seat of descending degeneration. Above the sections appear as Plate 68; Fig. 2 of that plate being from this case.

Fig. 3.—"*Compression Myelitis.*" A carcinoma of the third dorsal vertebra which caused a compression of the dorsal cord and resulting severe degeneration of fibres in all parts of the section.
The case was that of a woman, 42 years old, who began to be sick with severe neuralgic pains in the back, which was quickly followed by paresis of the lower extremities; these increased until she was bed-ridden and were associated with some disturbances of the bladder and of sensibility. After three months complete paraplegia, formation of contractures, and such severe pain that it could scarcely be controlled by morphine. Duration of disease six months.

Explanation of Plate 70.

SYRINGOMYELIA.

FIG. 1.—*Hydromyelia of the Upper Dorsal Cord.*
(Professor von Strümpell's Preparations.)

A man, 35 years old, hereditary tendencies, sickened with slow, increasing weakness and stiffness of the legs; soon after of the arms, with increase of the tendon reflexes; spasm of the muscles became aggravated; no muscular atrophy, sensibility normal.

Diagnosis: Spastic spinal paralysis.

Autopsy: Hydromyelia with degeneration of the pyramidal tracts.

The central canal in its entire extent was distended into a wide, irregular cavity filled with fluid (hydromyelus). Congenital?

FIGS. 2 AND 3.—*Sections through the Upper and Middle Cervical Portions of Another Case of Syringomyelia.*

The cavity formation extends partly in the anterior, partly in the posterior horns, entirely penetrating the latter. The clinical course was as follows: A man, about 40 years old, developed emaciation of the right hand, the right arm, and later of the left arm and hand; the muscles of the ball of the thumb, hypothenar eminence, and intcrossei showed the degenerative atrophy first. On the arm the pain and temperature sensibility were entirely lost, the touch sense normal. Later there occurred trophic disturbance in the hands, inflammatory necrotic process in the phalanges, finger nails, and in the joints.

In Fig. 2 a partial ascending degeneration of Goll's columns (s) is recognizable consequent to the involvement of the posterior columns by the cavity formation of Fig. 3.

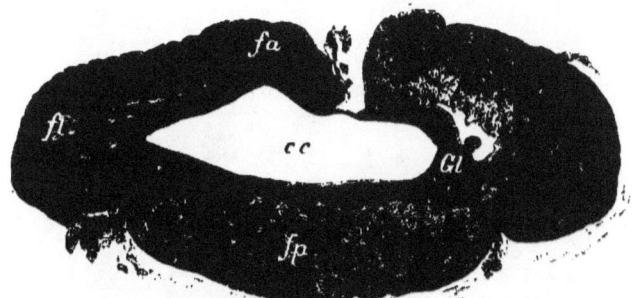

Fig. 1.

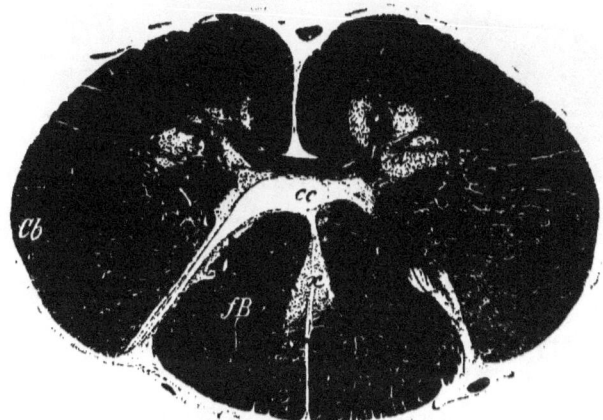

Fig. 2.

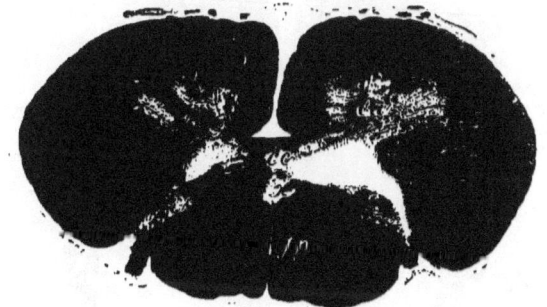

Fig. 3.

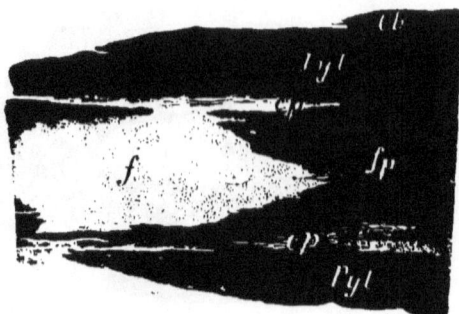

Fig. 1.

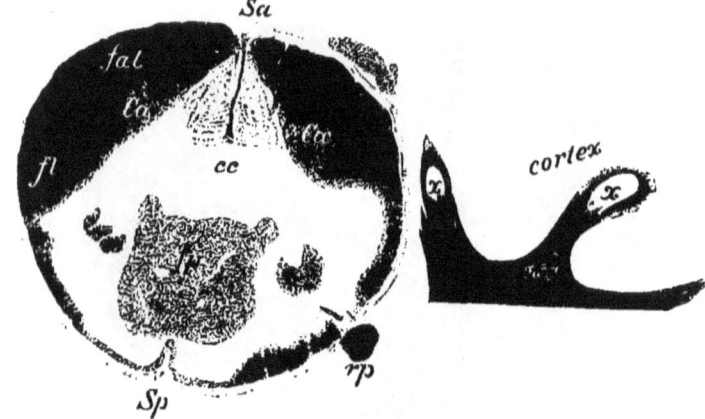

Fig. 2.   Fig. 3.

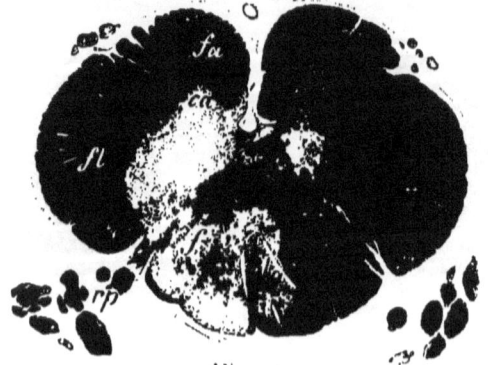

Fig. 4.

Explanation of Plate 71.

MULTIPLE CEREBRO-SPINAL SCLEROSIS.

A woman, 25 years old, remarked for several months that there was an increasing weakness of the hands; tremor and uncertainty on grasping anything (intentional tremor); to which later were added a stiffness and uncertainty of gait, nystagmus, scanning speech, optic atrophy (white, pale papillæ), increased tendon reflexes, sensibility normal, slight vesical disturbance. In three years the woman was completely bedridden and death followed a pneumonia.

Autopsy: In the brain and spinal cord were found regularly diffused sclerotic foci, within which the medullary sheath of the nerve fibres had become striated and disappeared, the axis cylinders remaining in part intact.

FIG. 1.—*Longitudinal Section through the Dorsal Cord* shows a sclerotic focus (f) in the posterior columns which have been cut through by the section.

FIGS. 2 AND 4.—*Transverse Sections of the Dorsal Cord.*

In 2 the entire transverse section is affected; only a part of the antero-lateral tracts is preserved. The gray substance of the anterior and posterior horns is diseased.

In 3 there is a small area to be seen in the posterior columns and in the anterior-posterior horns, especially on the left side.

FIG. 3.—*Section through the Cerebral Cortex* of the hemispheres.

In the radiating medullary substance are two small sclerotic foci (x).

Similar foci were found in the brain stem, pons, medulla, and in the cerebellum.

Explanation of Plate 72.

FIG. 1. —*Section through the Medulla of a Case of Chronic Bulbar Paralysis with Amyotrophic Lateral Sclerosis.*

A man, 30 years old, was taken sick with a slow, progressive atrophy and paralysis of the small muscles of both hands. Soon following there was distinct hindrance of speech (bulbar speech), the movements of the lips and tongue became slow, and the involved muscles continued to atrophy and there were lively fibrillary muscular contractions of the tongue. The gait was very slow and stiff. After two years there was complete atrophy and paralysis of the tongue and lips, paralysis of swallowing, speech impossible, arms entirely atrophic, and spastic paresis of the legs. Death from pneumonia of swallowing.

Preparations show : severe loss of cells and fibres in the hypoglossal nucleus (see Plate 56, 1), lightening in the pyramids. In the facial nucleus also a loss of cells.

FIG. 2. —*Section through the Lower Cervical Cord of a Case of Amyotrophic Lateral Sclerosis.*

The clinical course of the disease was very similar to the above-described case, except that in uncomplicated cases bulbar symptoms do not show themselves. Degenerative, slowly increasing atrophy of the muscles of the upper extremities beginning in the thenar and interossei groups ; spastic paresis of the legs ; sensibility and bladder intact. Increased, lively reflexes.

Preparations show : severe loss of cells and fibres in the anterior horns (diminution in size of the same) and degeneration of the pyramids on both sides.

FIG. 3.—*Section through the Anterior Horns of the Cervical Cord in a Case of Spinal Muscular Atrophy* (Strümpell).

A man, 35 years old (bad heredity), became sick with atrophy and paralysis of the thenar muscles and of the interossei. In the course of years the atrophy involved all the muscles of the arms, the shoulder girdle, the rib and neck muscles. Reaction of degeneration in the muscles, fibrillary twitchings, sensibility normal. No bulbar manifestations, no disturbance of gait.

Preparations : Striking lessening in size of the anterior horns. Marked cell disappearance in the anterior horns (see Plates 56, 6 and 55, 2). The anterior roots are also atrophic (compare their radiations in Plate 47, 1). In the anterior horns a close network of cicatricial glia tissue. Lateral columns normal.

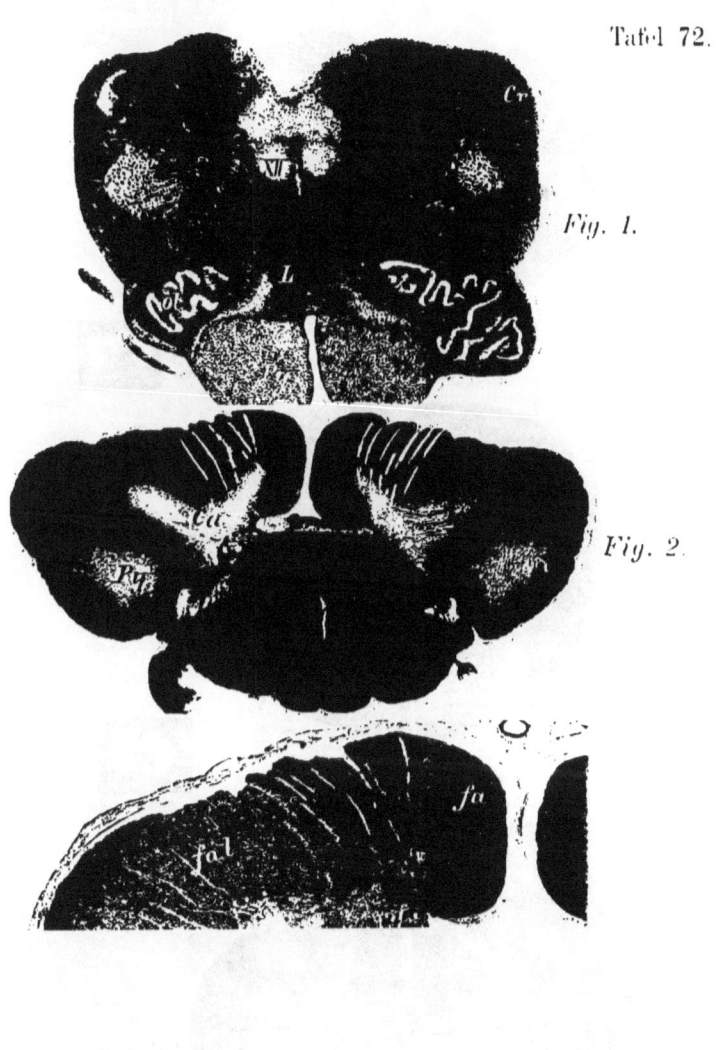

Tafel 72.

Fig. 1.

Fig. 2.

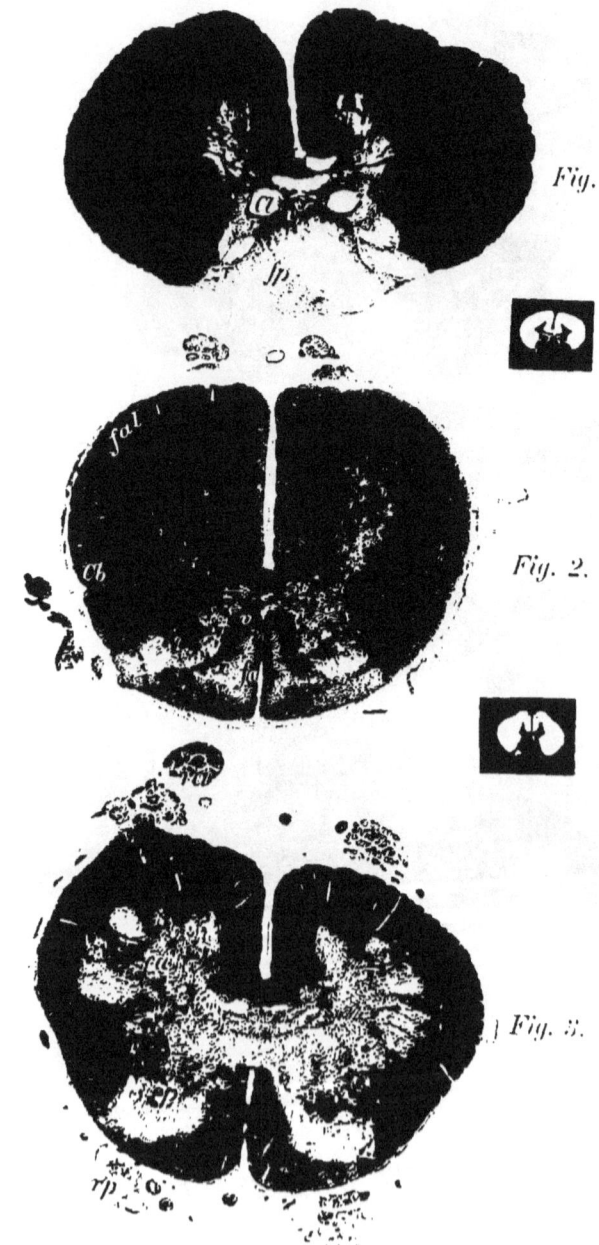

Explanation of Plate 73.

Tabes Dorsalis.

In tabes there is degeneration in part of the fibres of the peripheral sensory neurons, most frequently of the lower extremities, more seldom of the arms and face (trigeminus). Where the primary seat of the lesion is, it is not possible to say (neuron cells of the spinal ganglia?). The first to be diseased in all probability are certain reflex tracts (collaterals from the posterior roots), then the posterior roots (see Plate 77, 5) in their entire course, and the short fibres that pass into the posterior horns, Clarke's columns as well as the fibres of the long tracts (those that pass from the roots of the lumbar cord to the columns of Goll). The peripheral portion of the neurons is always degenerated (sensory cutaneous nerves). (See Plate 57, 4.)

Fig. 1.—*Section through the Lower Dorsal Cord.*

The posterior columns are enormously degenerated; the greater part of the ascending and descending fibres from the posterior roots are degenerated. The ventral field (v) of the posterior columns does not come from the posterior roots and it is in tabes invariably spared. The short tracts that radiate into the posterior horns from the posterior roots are entirely lost (compare normal section Plate 44, 1), Clarke's columns in consequence of the loss of fibres appear as clearly differentiated round white areas.

Fig. 2.—*Section through the Upper Lumbar Cord of Another Case.*

The degeneration is limited to symmetrical fields of both posterior columns, between which a small district is preserved (oval field, fo). In addition there is a so-called "border degeneration" in the borderland between the anterior and lateral columns. Principally the short descending tracts, which arise from the cells of the gray substance (central neurons), are degenerated.

Fig. 3.—*Section through the Lower Lumbar Region of a Third Case.*

Aside from the equal clearing to be seen in the posterior columns (ventral fields normal), the degeneration of the posterior roots (rp) is clearly recognizable, while the anterior (ra) are colored black and appear normal.

Explanation of Plate 74.

Tabes Dorsalis.

Fig. 1.—*Section through the Cervical Cord of the Case Plate 73, Fig. 1.*

The continuation of the long ascending posterior root fibres of the lumbar cord, forming the column of Goll, is totally degenerated. In Burdach's column there is likewise an area of lightening in the middle zone (long fibres from the dorsal and cervical cord roots). The medullary bridge at the apex of the posterior horns (*L*, column of Lissauer) which contains the most lateral fine root fibres is degenerated.

The preparations are from a man, 50 years old, who for six years complained of distressing sensitiveness in the region of the stomach, which later became a girdle feeling about the entire abdomen. During the last four years there was increasing uncertainty of gait, frequent attacks of neuralgiform pain in the legs (lancinating pains). Paræsthesiæ. In addition there was loss of the patellar reflex and of the pupillary light reflex, severe ataxia of locomotion, distinct disturbance of sensibility (delayed pain sense), disturbance of the bladder. Later there developed a chronic swelling in the left knee joint (tabic arthropathy), curved knees, and the ataxia became so severe that he was completely bed-ridden.

Fig. 2.—*Section through the Cervical Cord of a Case of so-called "High Tabes."*

In addition to the disease of the peripheral sensory neurons of the lower extremities, those for the upper extremities are likewise very much diseased, and it is thus that the entire posterior columns are degenerated in the cervical cord (Goll's and Burdach's columns). There is, furthermore, an asymmetrical disease of the lateral pyramidal tracts (slight lightening in the same). We have here, therefore, a combined tabes.

Fig. 3.—*Section through the Upper Cervical Cord. Case of Combined Tabes.*

In this case, also a "high" tabes, a portion of the posterior columns, the posterior outer field, is preserved. In addition there is a distinct bilateral degeneration of the lateral cerebellar tracts (tabes combiné). Clinically in this case there was striking muscular paresis and preservation of the pupillary reflex! It was nevertheless looked upon as a possible case of "pseudotabes."

Tafel 74.

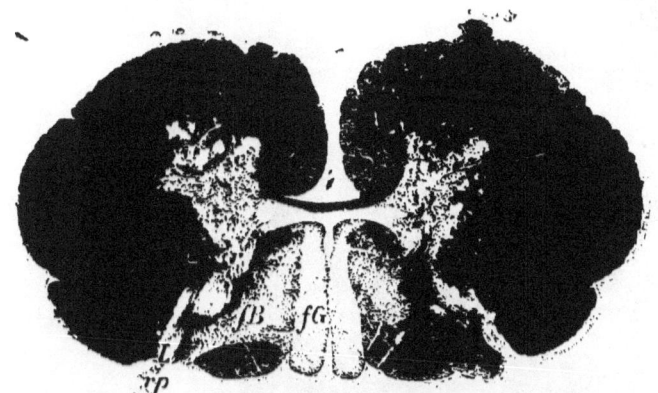

Fig. 1.

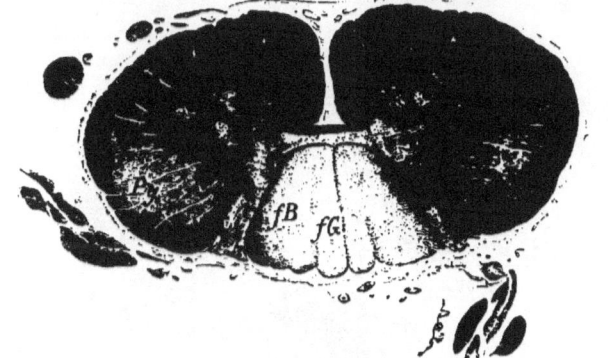

Fig. 2.

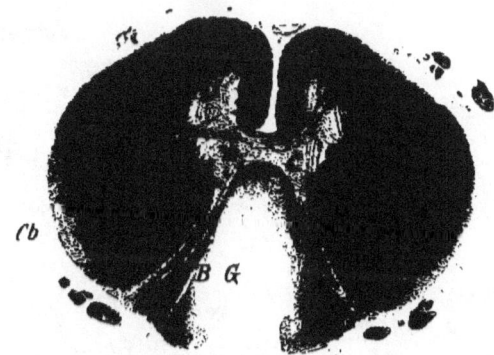

Fig. 3.

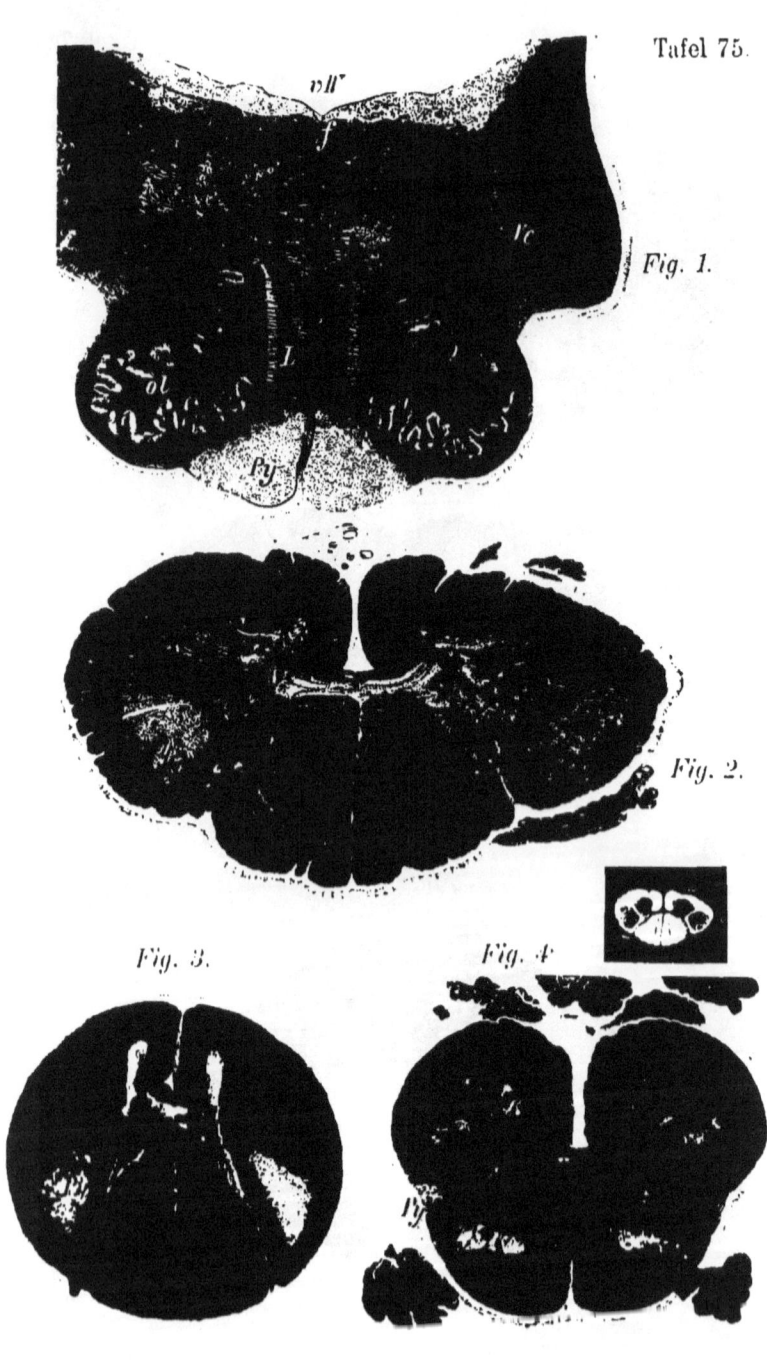

Tafel 75.

Fig. 1.

Fig. 2.

Fig. 3.   Fig. 4.

Explanation of Plate 75.

SPASTIC SPINAL PARALYSIS.

The preparations are from an extremely interesting case, which Professor Strümpell has published recently. It is the only case of purely spastic spinal paralysis which is clearly explained anatomically. (This form has been postulated on theoretical grounds for years by Charcot and Erb.) A woman, 63 years old, sick for about two years with gradually increasing weakness of the legs and arms. The paresis gradually increased to complete paralysis of the legs and arms, the muscles became rigid, allowed of but slight passive motion, spastic resistance, lively reflexes. No muscular atrophy, no disturbance of the bladder or of sensibility. Eventually (three years after) entire paralysis of the voluntary skeletal muscles (eye, face, masticatory, tongue, neck, arm, trunk, leg muscles). All the others normal. Anatomically there was found symmetrical degeneration of the pyramidal tracts in their entire course from the cortex of the brain through the internal capsule (see Plate 57, 6), the brain stem, the medulla, and spinal cord (central motorial neuron complex).

FIG. 1.—*Section through the Medulla.*
Both pyramids are degenerated, pale, a greater part of their nerve fibres have disappeared.

FIG. 2.—*Section through the Cervical Cord.*
Degeneration of both pyramidal tracts. In addition there is a slight cell atrophy in the anterior horns. (Approaching amyotrophic lateral sclerosis.)

FIG. 3.—*Section through the Middle Dorsal Cord.*
A lightening in the lateral pyramidal tracts.

FIG. 4.—*Section through the Lumbar Cord.*
Same condition as in 3. The diseased pyramidal tracts have approached the periphery.

Explanation of Plate 76.

THE COMBINED SYSTEM DISEASES.

Besides combined tabes there are other cases in which there occurs not only disease of neuron complexes which stand in anatomical or functional relationship, but in which simultaneous primary degeneration of neuron complexes of different function takes place. The point of origin of the degenerative processes (primary disease of the neuron cell? neuron process?) is not yet definitely known.

Clinically these cases present a slow, progressive, spastic paresis of the legs, arms, and also ataxia of the same, combined with more or less clearly demonstrable disturbance of sensibility and of the bladder. No muscular atrophy. Such a disease, for instance, is hereditary ataxia (see text). The following are sections from the spinal cord of such a case:

FIG. 1.—*Section through the Cervical Cord.*

Symmetrical disease of the lateral pyramidal and the anterior pyramidal tracts, and of the lateral cerebellar tracts of Goll's columns, and of the middle district of Burdach's columns.

FIG. 2.—*Section through the Lower Dorsal Cord.*

As in section 1, in the posterior columns the degeneration is less extensive and involves more the middle part of the posterior columns.

FIG. 3.—*Section through the Lumbar Cord.*

The lateral pyramidal tracts very severely degenerated; in the posterior column only a very small area of lightening.

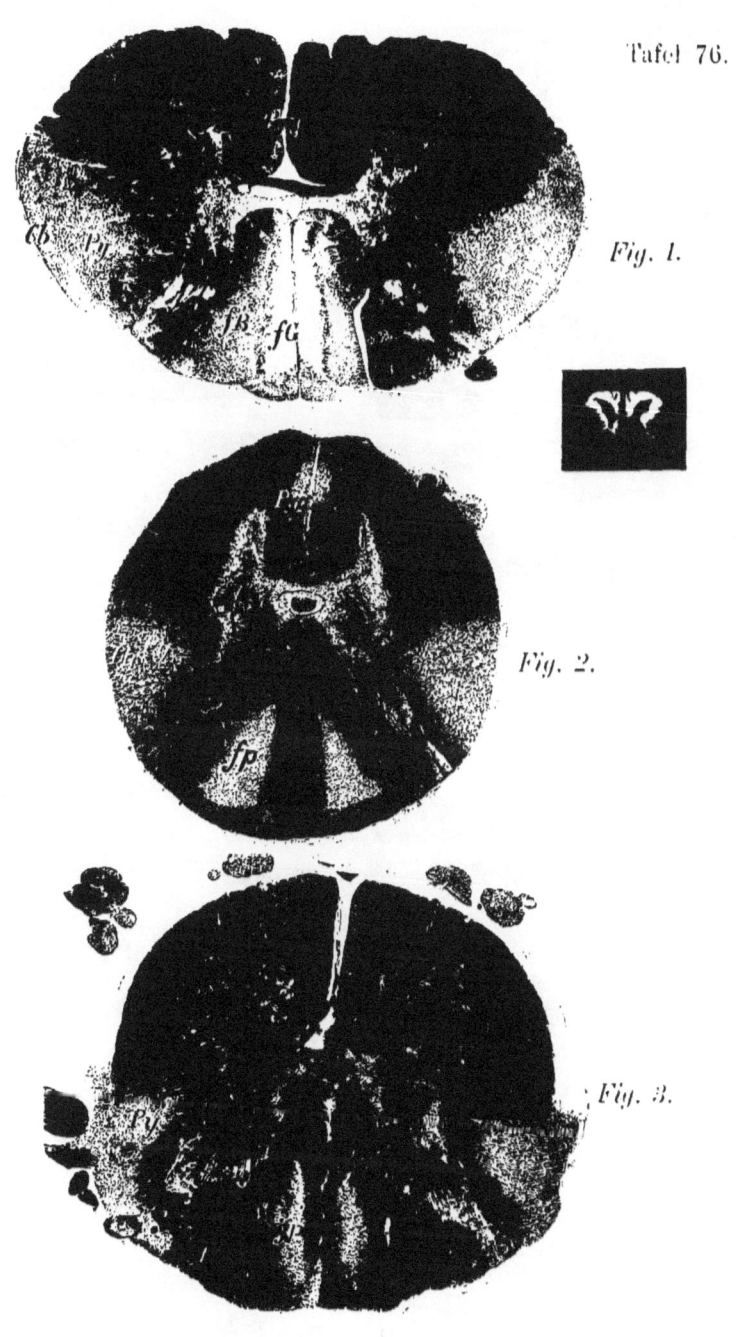

Tafel 76.

Fig. 1.

Fig. 2.

Fig. 3.

Tafel 77.

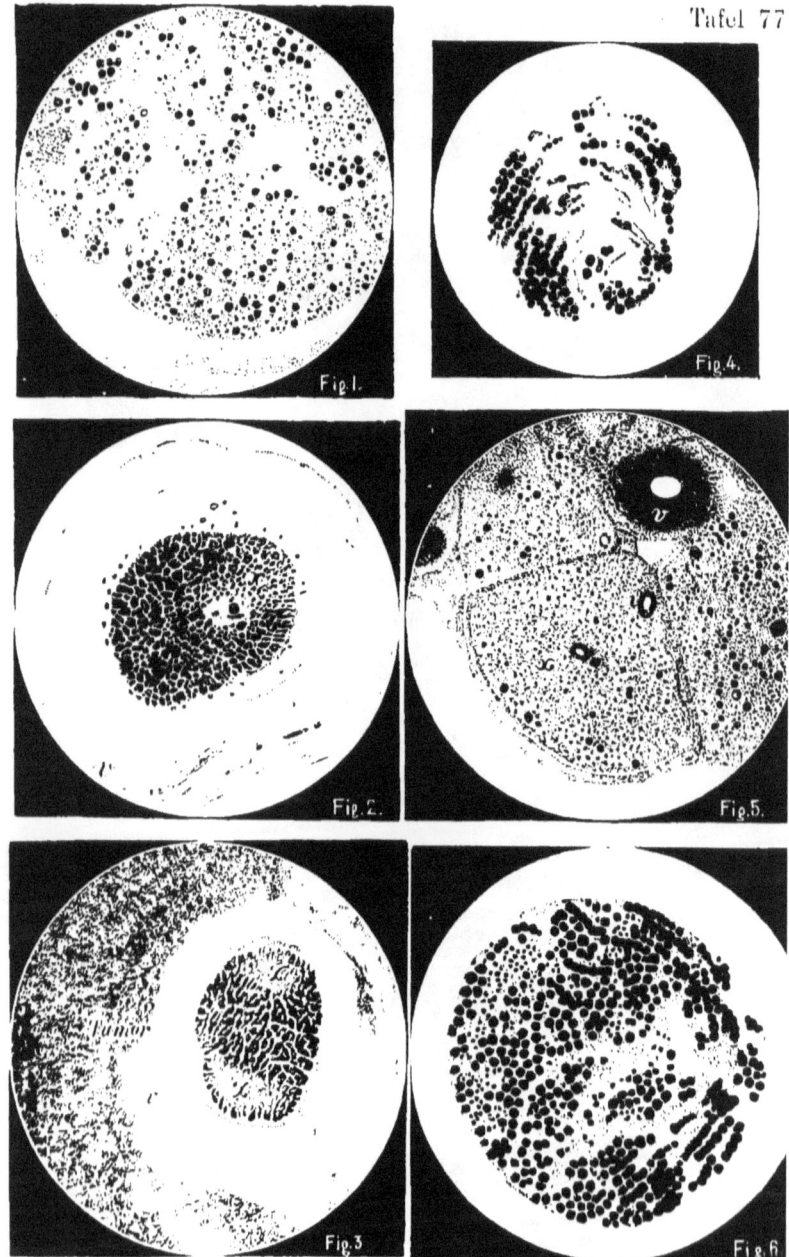

Explanation of Plate 77.

DEGENERATION OF PERIPHERAL NERVES.

FIG. 1.—*Ascending Secondary Degeneration in the Sciatic Nerve after Amputation of the Leg.*

A man fourteen years previously had suffered amputation of the leg. The figure shows the boundary of a nerve bundle, in which the larger part of the nerve fibres has disappeared, especially the coarser fibres are degenerated secondarily in an ascending direction (the result of loss of function caused primary neuron-cell disease in the lumbar cord?) ; the finer are somewhat better preserved. (Compare with the normal transverse section, Plate 46, 3.)

FIG. 2.—*Descending Degeneration of the Right Optic Nerve after Destruction of the Left Lateral Geniculate Body and the Right Corpora Quadrigemina.*

In the optic nerve pass fibres the greater part of which have their neuron cells in the retina alongside of those which have their cells in the so-called subcortical optic centres. These last degenerate in a descending direction, the others in an ascending. We recognize here a transparent sector-shaped area (x) in the optic nerve. (Compare with the normal optic nerve, Plate 46.)

FIG. 3.—*Compression Neuritis of the Optic Nerve.*

A case of acromegaly which had a malignant tumor of the hypophysis, which latter surrounded the optic nerve and caused partial degeneration (x, x). v, sheath of the optic nerve (dural sheath).

FIG. 4.—*A Motor Nerve Branch in Spinal Muscular Atrophy.* One readily sees the considerable loss of fibres widely diffused. (Belongs to the case from which Plate 72, 3 is prepared.)

FIG. 5.—*Posterior Root from the Lumbar Cord in a Case of Tabes.*

(Belongs to the case, Plate 73, 3.) The bundle which is represented (x) is from the posterior root and contains scarcely ten normal fibres; to the right of it the fibres are somewhat better preserved. V, blood-vessel.

FIG. 6.—*Bundles of the Peroneal Nerve from a Case of Neural Muscular Atrophy.* One recognizes the loss of nerve fibres in focal areas, especially in the district of the middle bundle. Involvement of the spinal cord is not excluded (motor anterior-horn cells?).

Explanation of Plate 78.

MULTIPLE NEURITIS.

FIGS. 1 AND 2.—*Preparations from Severe Cases of Alcoholic Neuritis (1 from the Crural Nerve, 2 from the Sciatic).*
A man, 40 years old, heavy drinker, was taken sick rather acutely with general weakness of the extremities which rapidly became complete paralysis. Great sensitiveness on pressure over the nerve trunks, loss of patellar reflexes, disturbances of sensibility, rapidly developing muscular atrophy. Strychnine injections made his condition worse. Presence of psychical disturbance. Death after two and one-half months.

An advanced degree of destruction of nerve fibres in all the peripheral nerves was found. In many bundles but from six to twelve (see Fig. 2) normal transverse sections of nerves are to be seen (compare Plate 46, 2). This degeneration extended up as far as the anterior roots of the spinal cord.

FIG. 5.—*Transverse Section of the Cervical Cord of this Case.*
One recognizes the uncommonly distinct atrophy of the anterior roots in their passage through the antero-lateral tracts (ra); there can be seen also in the anterior horns a perceptible lightening, particularly in the middle portion (m), which is secondary to the degeneration of the anterior roots which enter here.

In Goll's columns slight ascending degeneration (long fibres of the posterior roots).

FIG. 3.—*Post-Diphtheritic Neuritis.*
Cross section of a bundle from the crural nerve showing high degree of fibre degeneration. Taken from a 15-year-old boy, who, after convalescence from a severe pharyngeal diphtheria, developed acute pharyngeal paralysis and shortly afterward paralysis of the extremities. Patellar reflexes were lost within ten days, slight pain and disturbances of sensibility. Death from paralysis of the vagus.

FIG. 4.—*Longitudinal Section of a Nerve Bundle from the Sciatic Nerve of a Patient with Infectious Multiple Neuritis (Landry's Paralysis).*
In a part of the degenerated nerve fibres are seen the products of cell destruction (little balls of myelin) which have not been transported away. The nerve fibres are found in all conceivable stages of degeneration.

The preparation is from a case in which the cause of the disease was unfathomable. A woman, 30 years old, sudden severe sickness (fever?), complete paralysis first of the legs, then soon after of the arms (ascending paralysis). The patient soon became unconscious, was very much neglected for the first two day, but death came while under treatment by a physician. (Spinal cord not examined.)

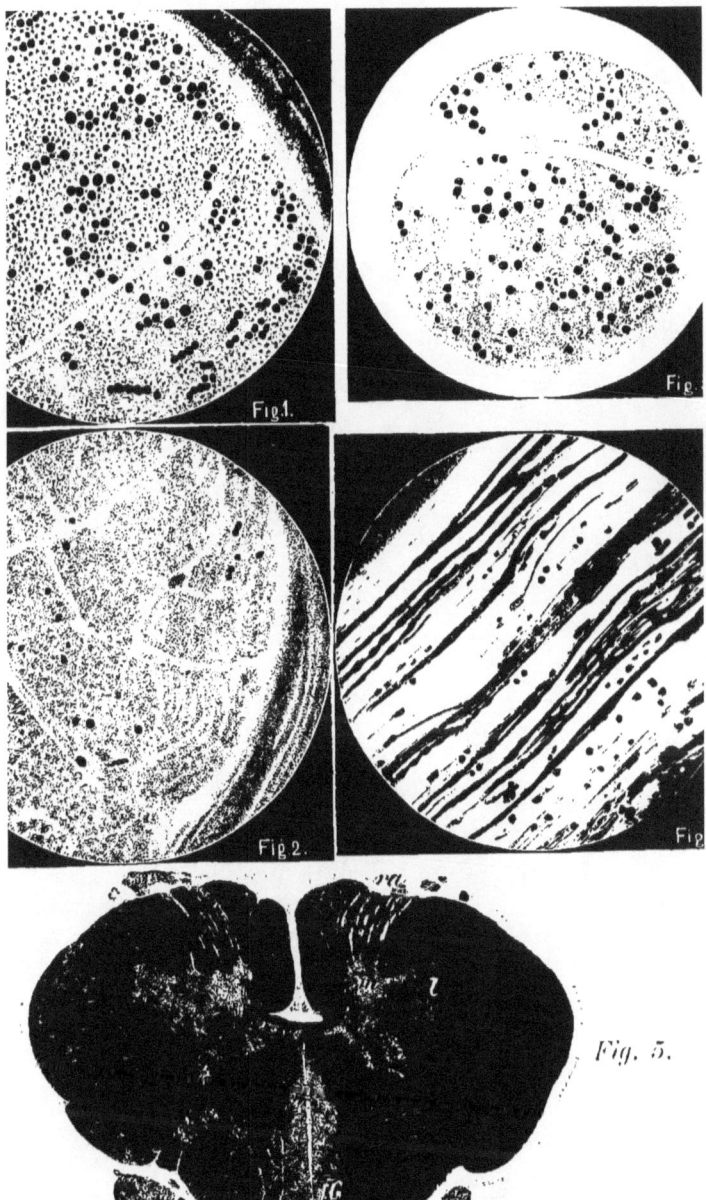

SECTION I.

Morphology of the Nervous System.

(Plates 1 to 10.)

THE nervous system of man consists of (1), the central organ (brain—spinal cord) with the peripheral nerves that spring from them; and (2), the sympathetic nervous system, consisting of a chain of sympathetic ganglia and their fibre ramifications.

These two divisions are closely connected with each other by means of numerous nerve tracts.

The brain and spinal cord lie in a bony capsule or protective framework, the cranial cavity and the spinal canal, from out of which the peripheral nerves pass to the soft parts. The sympathetic chain of ganglia lies on both sides of the vertebral column and immediately in front of it.

The brain and spinal cord are surrounded by three membranous envelopes, the meninges. The outer one, lying immediately beneath the bone and serving as periosteum in the cranial acvity, is called the dura mater. It is a firm, dense, fibrous membrane. It surrounds the brain and spinal cord like a sac and covers also the outgoing nerves as far as their canals of exit where it ends, blending intimately with the margins of the foramina through which the nerves pass. In the dura course the broad collective channels for the venous blood of the brain, blood which flows into the internal jugular vein from the sinus venosi (sinus falciform, superior, inferior, cavernosus, petrosus, transversus, etc.).

Prolongations of the dura pass into the interior of the skull cavity. The vertical portion of the dura that passes from above downward in the sagittal median line is adherent anteriorly to the crista galli, great falciform process (sickle process), which divides the skull cavity in the upper half into two parts. The transversely coursing tentorium serves as a roof for the posterior cavity of the skull.

The dura surrounds the brain and spinal cord loosely, and beneath it is found a space filled with lymph, which is known as the subdural lymph space.

The second envelope is a thin, cobweb-like covering, the arachnoid. It is in close apposition to the brain, but less so to the spinal cord; it bridges over the depressions in the brain without sinking into them. The Pacchionian bodies on each side of the falciform process are structurally connective-tissue constituents of the arachnoid. Under the arachnoid is the subarachnoidal lymph space, containing cerebro-spinal fluid, consisting of numerous compartments which communicate one with another. The most internal envelope, the pia mater, in conjunction with the arachnoid form what are usually called the "meninges." The pia dips into all the fissures and depressions of the brain in close contact with the nerve substance. The inward prolongations into the depressions are known as tela choroidea. It transports the proficient vascular supply of the brain and spinal cord.

The brain consists of two hemispheres and the

brain stem. The direct continuation of the latter portion is the spinal cord. The latter passes nearly perpendicular to the position of the former, owing to the fact that the brain stem below bends downward at an angle of nearly 90°.

The weight of the fully developed brain varies from 1,300 to 1,400 grammes.

The two hemispheres, which are connected one with another only on their median surfaces, consist of a superficial enveloping gray substance, the cortical substance, cortex cerebri, and of a central white mass beneath the cortex, the medullary substance.

The cortex does not spread itself smoothly over the medullary substance of the hemispheres; on the contrary, following the formation of the numerous sulci it is arranged in worm-like gyri or convolutions. It is to this arrangement that the greater area of the gray matter is due.

Of the convolutions, one part is constant while the other depends on individual variation. Each hemisphere is made up of various lobes; each lobe comprises several cortical convolutions and one portion of medullary substance.

In the anterior part of the cranial cavity lie the frontal lobes, constituting the anterior pole of the brain. They have two principal fissures, superior and inferior frontal sulci, which divide the frontal lobes into three convolutions, the superior, the middle, and the inferior frontal gyri.

Behind the frontal lobes, somewhat posterior to the middle of each hemisphere, passing from posteriorly above to anteriorly below, is a deep sulcus, the central fissure of Rolando. It separates the two central convolutions, the anterior central gyrus and the posterior central gyrus. Behind the central convolutions lies the parietal lobe, which is divided by the interparietal sulcus into the superior and inferior parietal lobes.

On these lobes border the occipital lobes, constituting the posterior pole of the brain. They are subdivided like the frontal lobes into three convolutions —the superior, middle, and inferior occipital gyri. The basilar surface of the occipital lobes rests upon the tentorium. The temporal lobes lie toward the middle of the cranial cavity. They have three fissures, the superior, middle, and inferior temporal sulci; and three convolutions, the superior, middle, and inferior temporal gyri. The temporal lobes in front and above are separated from the frontal and central convolutions by a very deep broad fissure, the fissure of Sylvius, which posteriorly passes directly into the inferior parietal convolutions. It thus gives rise to three arciform convolutions which in the inferior parietal lobe are known, enumerating from in front backward, as the supramarginal gyrus, the angular gyrus, and the preoccipital gyrus. At the bottom of the Sylvian fissure lies a lobe covered over by the neighboring convolutions, especially by the temporal and central lobes. This lobe is composed of a number of small convolutions and is known as the island of Reil.

The convolutions that have thus far been considered are on the external convex surface of the hemispheres. On the median surface are found in front the upper continuation of the superior frontal convolutions. The central convolutions unite in the paracentral lobes and the parietal lobes continue as the præcuneus. The latter is separated from the median surface of the occipital lobe by the parieto-occipital fissure. The median surface of the occipital lobe is divided into the cuneus above and lingual lobe below, separated by the calcarine fissure.

On the under surface of the temporal lobes, the median portion of the occipito-temporal gyrus borders on the inferior temporal gyrus, and separated by the occipito-temporal fisssure lies the inferior marginal convolution, the gyrus hippocampus, which ends anteriorly as the uncus. On the base of the frontal lobes, the fissures of importance are the straight sulcus and laterally to this the tri-radiate sulcus, to which belongs the middle frontal gyrus. Aside from the above-mentioned convolutions on the median surface of the hemisphere there are other relationships to be considered. The cortex terminates superiorly as the gyrus fornicatus and below as the gyrus hippocampus (superior and inferior marginal convolutions). Beneath the superior marginal convolution the white substance breaks through both parts, unites and constitutes the corpus callosum.

Above the inferior marginal convolution the white medullary substance passes out of the two hemispheres converging as in the corpus callosum and uniting to form a part of the brain stem at its beginning (crusta).

Between the corpus callosum and the cerebral peduncles are a number of formations and spaces which will be considered later on. Here it only remains to call attention to the fact that the entire middle portion named, bordering upon the marginal convolution, has no longer a covering of cortical substance.

The corpus callosum, which connects the two hemispheres, is made up of thick layers of white substance. Its larger, middle portion passes in a nearly horizontal direction. The anterior part, the knee of the corpus callosum, bends basalward and just before its termination somewhat backward. The posterior end thickens into a cushion-like prominence and is known as the splenium of the corpus callosum. In the interior of each hemisphere is found a mass of white medullated substance, constituting the core of the various lobes of the hemispheres as previously described. In its upper half, up to the level of the corpus callosum, this mass of white substance is formed by the radiation of fibres from this structure. As shown in cross section at this level, it is called the centrum semiovale of Vieussens. In its basal portion there is seen embedded in the medullary substance a grayish-red mass the size of a goose egg. This constitutes the root ganglion of the brain—the corpus striatum. It is divided into the lenticular nucleus, situated laterally; and the caudate nucleus, situated mesially. This separation is produced by a white tract passing through this region from the cortex of the various cerebral lobes. It emerges from the base of each hemisphere as the crusta.

In the interior of each hemisphere is a system of cavities filled with lymph, the lateral ventricle (ventricul. lateral.). The roof of each lateral ventricle is formed by the median half of the centrum semiovale of that side, and because of this circumstance it is here known as the tegmentum of the ventricle. The basal ganglia, under the name of the corpus striatum, pass out of the medullary substance as a free surface and project outward and below into the lateral ventricles.

The part of the lateral ventricle that is situated between the under surface of the lateral portions of the corpus callosum above and the basal ganglia below is known as the middle cell (cellula media). The ventricle from here extends by its anterior horn (cornu anterioris) into the anterior lobe, by its posterior horn (cornu posterioris) into the occipital lobe, and by its inferior horn (cornu inferioris) into the temporal lobe. The anterior and posterior horns are entirely surrounded by the white substance of the hemisphere. The inferior horn, on the other hand, is not closed in the median line (inferior aperture). This results from an involution and contraction of the inferior marginal convolution, which forms in consequence a cushion or pad-like projection into the inferior horn, known as the cornu Ammonis. The exact end of the cortex of the gyrus hippocampus forms the small gyrus dentatus (fascia Tarini). This relationship is well seen in the section, plate 27.

From the free end of the cornu Ammonis arises the medullated fimbria, forming the termination of the white substance of the gyrus hippocampus. It passes along with the superficially indentated Ammon's horn "digitationes corpus Ammonis" from the anterior end of the inferior horn backward and upward. At the point of divergence of the inferior horn from the cella media, the fimbria separates from the cornu Ammonis, uniting with its fellow of the opposite side as the ascending pillars of the fornix "crus ascendens fornicis," converging toward the inferior surface of the posterior portion of the corpus callosum, and thus forms in the middle a triangular space, the psalterium. The united fornix passes beneath the corpus callosum and unites with it some distance forward. It divides, however, before it reaches the genu of the corpus collosum and again descends as the two descending pillars of the fornix. At the base the descending crura of the fornix turn backward and terminate in the bilateral mammillary bodies.

Between the knee of the corpus callosum and the descending pillars of the fornix there remains in the middle, in consequence of the relations already described, a free space which is filled out by two small partitions, the septa pellucida, with a small cavity lying between, the ventricle of the septa pellucida. To the side of each septum lies the lateral ventricle beneath the corpus callosum.

From the floor of each lateral ventricle, passing up abruptly out of the medullated substance of the hemisphere, there projects into the ventricle the free upper surface of the corpus striatum. Its anterior broader portion, extending toward the anterior horn, is called the head of the corpus striatum (in section nucleus caudatus). From the head there passes backward and outward the small tail of the latter, which bends downward at the place of departure of the inferior horn and finally continues forward from here as a portion of the roof of the inferior horn.

In a direction median to the corpus striatum runs a small white band parallel with it, the stria cornea. It separates the corpus striatum from a structure that rises posterior and median to the corpus striatum, the optic thalamus, a mass which on cross section appears like the basal ganglion of a reddish-gray color.

The two thalami approach each other posteriorly and somewhat toward the median line. The cavity between them is the third or middle ventricle. The lateral ventricles of the hemispheres, particularly the cella media, are in direct connection with the third ventricle by means of a small aperture immediately behind the descending pillars of the fornix, the foramen of Munro. The fornix with the corpus callosum lying above it would constitute the roof of the third ventricle were it not that from the pia mater beneath the splenium corporis callosi (transverse fissure) a process of this membrane projects into the third ventricle, which in this locality is open. This vascular projection of the pia passes beneath the fornix covering the third ventricle and spreads out thinly anteriorly as the tela choroidea superior. It carries in its central portions two band-like, granular-looking structures, plexus chorioidei medii. These pass through the foramen of Munro out of the third ventricle into the lateral ventricles as the plexus chorioidei laterales. In the lateral ventricles they pass backward resting on the lateral border of the thalamus, finally turning about into the inferior horn. Here they hang together with the pia, which has pressed in through the aperture in the inferior horn.

As described above, the white substance of the hemispheres is compressed in the middle portion of each hemisphere owing to the gray masses of the basal ganglia and of the optic thalamus, which take their origin in that locality (the middle part) and increase in extent toward the base of the hemispheres. The medullary tracts which pass out of the centrum semiovale deeper down break through the basal ganglion as the anterior limb of the internal capsule and thereby divide the lateral lenticular nucleus in the direction of the insular lobe from the nucleus caudatus. Another portion of the white substance of the hemispheres passes downward as the posterior limb of the internal capsule between the lenticular nucleus and the optic thalamus. Both limbs come into contact at the knee of the internal capsule. The constituents of the medullary substance of the internal capsule pass in part into the basal ganglia and optic thalamus, disappearing therein; in part they continue in the internal capsule further toward the basal region and make their exit as the crusta from the hemispheres at the base of the brain.

The wedge-shaped lenticular nucleus which forces itself between the internal capsule and the lobes of the island of Reil divides into several parts, a lateral, larger putamen, and several internal portions, which lie in a median direction. These are from a developmental point of view to be distinguished from the putamen. They are called globus pallidus. The putamen and the nucleus caudatus together comprise the actual basal ganglion. Lateral to the putamen the white substance of the hemisphere is continued inferiorly as the narrow external capsule. Between this and the medullated substance of the lobe of the insula lies another small gray mass running parallel to the island, the claustrum. The medullary substance of the hemispheres, which, as has already been described, is compressed into small bands (capsula interna and externa), increases again in extent below the lenticular nucleus, especially in the posterior direction where it constitutes the medullated substance of the temporal lobe. Another small, round, gray mass, the nucleus amygdalæ, is found embedded in the medullary substance at the base of the brain below the anterior portion of the lenticular nucleus between the frontal and temporal lobes.

After removal of the corpus callosum with the fornix and tela choroidea the thalamus opticus lies with its upper and median surface free. It exhibits anteriorly a little elevation, tuberculum anterior, and expands posteriorly as the pulvinar.

The median surfaces of the thalami constitute the lateral walls of the third ventricle. Passing along the upper inner edge of the thalami extends the small tenia medullaris (semicircularis), rising up from below, converging posteriorly as the pedunculus conarii, and decussating with the peduncule of the other side. Laterally to the place of decussation lies on both sides the small ganglion habenulæ. The decussation has no connection with the pineal gland, a structure of about the size of a hazelnut (epiphysis conarium, glandula pinealis). Three short commissural bands run transversely across the space of the third ventricle between the two sides. Immediately in front of the descending peduncles of the fornix, deeply placed, is the large white anterior commissure. Through the middle of the third ventricle passes the soft lacerable gray commissure, and immediately under and in front of the decussating pedunculi conarii is found the posterior commissure.

Below the thalamus, on both sides of the brain, lies

a region into which a portion of the medullary substance enters from the lenticular nucleus, optic thalamus, and internal capsule, the so-called subthalamic region. In it lies the oval subthalamic body and the beginning of the red nucleus, the main portion of which is situated more caudad in the brain stem.

The third ventricle terminates anteriorly between the two descending, somewhat diverging limbs of the fornix. Inferiorly it narrows like a funnel to the infundibulum, the termination of which is made by the hypophysis cerebri in the sella turcica of the sphenoid bone. Posteriorly the ventricle becomes shallower in consequence of the median surfaces of the thalamus, whose lower portions converge gradually and continue beneath the posterior commissure. This closes in the ventricle from above and forms a narrow canal, the aqueduct of Sylvius. While the lower halves of the thalamus unite, the upper halves are pressed apart in their caudal portion by the corpora quadrigemina. These bodies take their origin behind the posterior commissure. They are divided into the two anterior and the two posterior portions—the anterior and posterior quadrigeminal bodies. Laterally from the corpora quadrigemina pass the two arms (brachia), the brachium anterior and posterior, to the two geniculate bodies, the lateral and median, which lie between the corpora quadrigemina and the pulvinar. From the lateral geniculate body the white optic tract passes toward the base of the brain.

In front of the broad pontal fibres the white cerebral peduncles converge. In the space between the crura cerebri is the posterior perforated substance, in front of which are two white bodies about the size of peas, the mammillary bodies or corpora candicantia. In front of these is the continuation of the third ventricle, the infundibulum with the hypophysis. In front of the latter is the chiasm of the optic nerves, made up of the continuations forward of the optic tracts of the two sides after they have rounded the cerebral peduncles. Beneath the corpora quadrigemina the system of cavities continues as the aqueduct, the walls of which are composed of the central ventricular gray substance which line the third ventricle, covering the surface of the thalami.

Beneath the corpora quadrigemina lies the tegmental region, which is formed from the subthalamic region by the junction of the thalami. Basal to the tegmentum are found the converging crura cerebri, which emerge from the two hemispheres and which seem to be the continuation of the white substance of the internal capsule. Beneath the tegmentum they form the crusta. Between the tegmentum and the pes lies a brownish, half-moon-shaped mass, the substantia nigra. Behind the corpora quadrigemina is the cerebellum, thrusting itself, covered by the tentorium, beneath the occipital lobes, into the posterior cranial fossa. In the median line and between the two cerebellar hemispheres lies the worm (vermis).

Beneath the corpora quadrigemina the brachium conjunctivum (processus cerebelli ad corpora quadrigemina) passes from the tegmental region into the cerebellum. Between these the thin anterior medullary velum forms with the lingula, the direct passage to the vermis (the valve of Vieussens).

The aqueduct passes beneath the valve of Vieussens and spreads out into the fourth ventricle bounded by the diverging superior cerebellar peduncles. The accompanying Fig. 1 illustrates the connections of the ventricular system.

The floor of the fourth ventricle constitutes a part of the fossa rhomboidalis. Beneath this are continued the component parts of the medulla oblongata. They are continued from the tegmentum and pes of the crura cerebri.

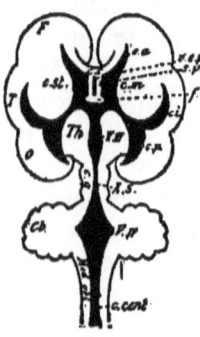

Fig. 1.

The roof of the fourth ventricle is formed by the under portion of the worm (vermis inferior). A process of the pia, known as the tela choroidea inferior, enters into the ventricle from behind.

The hemispheres of the cerebellum are composed, like those of the cerebrum, of the superficial gray cortex, and the central white medullary substance. They are divided into innumerable small, generally parallel convolutions which are arranged in single lobes. Every lobe is in continuous connection with the corresponding portion of the other hemisphere by means of a part of the vermis.

On the upper surface are two hemisphere lobes, the lobus superior, anterior and posterior (semilunaris), both united by the superior vermis (lobus centralis, monticulus, folia cacuminis).

On the under surface are found, enumerated from in front, the following hemisphere lobes: flocculus, tonsilla, lobus cuneiformis, posterior inferior lobe (semilunaris). The inferior vermis supplies the connection of these lobes (nodulus, uvula, pyramis, commissura brevis, are the several components in respective order). In the interior of the medullated area of the cerebellar hemispheres lies a gray puckered substance, the corpus dentatum cerebelli, and in the medullary area of the vermis the nucleus tegmenti (roof nucleus). Other smaller gray masses, emboliformis nucleus and globosus, lie between these two.

From the medullary substance of the cerebellum the superior cerebellar peduncles already described converge anteriorly, while inferiorly passes the processus cerebelli ad pontem, composed of white medullated bundles which converge posteriorly over the descending cerebral peduncles and at the base of the brain form the pons (pons Varolii). In a posterior direction run convergingly the processus cerebelli ad medullam oblongatam or the restiform bodies. The superior cerebellar peduncles form the lateral boundaries of the anterior half of the rhomboid fossa, the restiform bodies form the posterior half. The shape of the rhomboid fossa is determined by the convergence of these two medullary tracts as they pass from the cerebellum.

The fourth ventricle is covered with a layer of gray substance (central gray matter), and in the middle of the cavity the medullary white substance of the striæ acusticæ pass transversely toward the two sides at an obtuse angle.

Beneath the gray matter lies the continuation of the constituents of the tegmentum; beneath this, the continuation of the crusta encircled and divided by the transverse pontal fibres of the cerebellum. In addition to this there is a number of small gray masses (nuclei) lying between all of these interlaced medullary pathways. All of these together with the nerves that pass out constitute the medulla oblongata. On the posterior border of the pons, at the base of the brain, pass the pyramids, the continuation of the crusta. Laterally to the pyramids in the medulla are the olivary bodies which on cross section appear as variously involuted puckered gray leaves.

Dorsally to the passage of the pyramids the fourth ventricle becomes narrower, owing to the converging restiform bodies which unite at an acute, angle the calamus scriptorius. At the most posterior portion of the angle the ventricle sinks deeper into the substance of the medulla and soon disappears entirely from the surface, lying toward the centre of the medulla as the central canal and continuing in this position throughout the entire spinal cord.

The medullated bundles of the restiform bodies, shortly before their junction, form into a median bundle, the funiculus gracilis, or Goll's column, and a lateral bundle, the funiculus cuneatus, or column of Burdach. Both of these continue seemingly uninterrupted throughout the cord as the posterior columns. Ventrally the pyramids approach one another at the surface, while deeper in the medulla they may be seen to constitute the motor crossing (decussation of the pyramids). In conjunction with the most ventral portion of the restiform bodies, which does not go to the posterior columns, these crossed pyramidal fibres form the lateral columns or tracts of the spinal cord (funiculus lateralis). The anterior columns (funiculus anterior) arise from the parts that lie under the pyramids and which come into view after the disappearance of the latter.

Around the central canal, in the centre of the spinal cord, there are grouped together as a continuation of the gray substance of the ventricles with its nuclear masses, the anterior and posterior horns. The gray substance is, therefore, like the ventricles, pushed toward the centre, while the white substance lies on the periphery. The medulla oblongata is thus transformed into spinal cord.

The spinal cord consists of a round white bundle about the size of the little finger. It is enclosed in its membranes and extends in the spinal canal as far as the upper border of the second lumbar vertebra. It gradually diminishes in size as it passes downward, but there are developed in its course two enlargements, one in the cervical and one in the lumbar portion. These enlargements result from a marked increase in the central gray matter of the anterior and posterior horns. Between these two enlargements lies the dorsal portion of the cord. The lower termination of the cord is known as the conus medullaris. The three columns, the anterior, lateral, and posterior, pass, as portrayed above, circumscribing the gray substance of the anterior and posterior horns. In the conus medullaris terminate the columns and horns of the cord. In the conus the lateral column is the most, the anterior column the least prominent. In this portion of the cord the gray substance predominates over the white; in the cervical cord, etc., the reverse is the case.

Between the two anterior columns runs the deep anterior longitudinal fissure. Between the two posterior columns is the narrow posterior longitudinal fissure. From the stem of the brain and from the spinal cord the peripheral nerves pass out. They are the continuation of medullary columns gathered together in bundles.

The nerves are divided into the twelve cranial, and the thirty-one pairs of spinal nerves. At the base of the brain the twelve cranial nerves, up to the fourth, take their origin in the following order:

I. The olfactory nerve. It is composed of the small numerous nervi olfactorii which pass through the lamina cribrosa of the ethmoid bone. These enter into the olfactory bulb, which lies in the straight sulcus (sulcus rectus) of the frontal lobe. The bulb passes posteriorly as the olfactory tract and terminates laterally in front of the chiasm in the trigonum olfactorium. The olfactory tract, etc., in man comprises an undeveloped portion of the brain, but in the lower animals it is for the most part developed into a very considerable cranial lobe.

II. Optic nerve. Both nerves pass from the eyeballs and converge to the chiasm, undergo partial decussation (nasal, most considerable bundle) and pass as the optic tract over the cerebral peduncles farther upward as far as the lateral geniculate bodies in which they seemingly disappear.

III. Motor oculi. It passes beneath the quadrigeminal bodies in many bundles, through the substance of the tegmentum and reaches the surface mesal to the peduncle at the anterior border of the pons.

IV. Trochlear nerve. It passes dorsally behind the posterior quadrigeminal bodies, totally decussates in the velum medullare, passes toward the base and then over the peduncles farther forward.

V. Trigeminal nerve. It passes out with an anterior (motor) and a posterior (sensory) root through the lateral fibres of the pons. The posterior root forms the Gasserian ganglion. Then the nerve divides into its three branches.

VI. Abducens nerve. It arises from the floor of the fourth ventricle, passes through the medulla, and passes out at the posterior border of the pons lateral to the pyramids.

VII. Facial nerve. It emerges in company with the following nerve externally from the olive at the posterior border of the pons.

VIII. Auditory nerve. It is composed of two roots, the vestibular nerve (anterior median root) and the cochlear nerve (from the cochlea) (posterior lateral root) and emerges with the facial.

IX. Glosso-pharyngeal nerve.

X. Vagus nerve. Both of these pass out together ventral to the restiform bodies in many roots. The first forms with others the petrosal ganglion; the second the jugular ganglion.

XI. The spinal accessory nerve. It emerges by many roots that pass out through the lateral columns from the upper cervical cord and medulla. Its inner branch goes to the pneumogastric.

XII. Hypoglossal nerve. It passes out in many bundles between the pyramids and the olive.

The spinal nerves are divided into the eight cervical, the twelve dorsal, the five lumbar, the five sacral, and the one (seldom two) coccygeal. Each arises from an individual spinal cord segment; consequently from the eight cervical, twelve dorsal, five lumbar segments, etc. Every nerve emerges from its spinal segment in two separate portions—"roots." Between the anterior and lateral columns emerge the anterior motor root (radix anterior); between the lateral and posterior columns the posterior (sensory) roots (radix posterior). Both roots unite into one nerve trunk. Before uniting with the anterior root the posterior root forms the intervertebral ganglion, a knuckle-shaped swelling in the intervertebral foramina. The nerves for the neck, the extremities, etc., comprise, before they pass into the soft parts through their numerous anastomoses, the nerve plexuses (plexus cervicalis, brachialis, lumbalis, sacralis, etc.). In the strict sense of the word the peripheral nerves arise primarily from the plexuses. Consequently the peripheral nerves are composed of mixed fibres from the anterior and posterior roots, not only of a single spinal segmental level, but of different levels. (For further details see Plate 23.) The nerve roots for the lower extremities, in consequence of their level of origin (level of the first lumbar vertebra, see Plate 23) run for some distance together with the conus medullaris before their departure through the intervertebral foramina, and below the conus constitute the cauda equina. The chain of the sympathetic divides into a series of small prominences (ganglia) which from their beginning at the neck (cervix) (ganglion supremum, medium, imum) are situated at the anterior border of the vertebral column. In front of each vertebra lies a ganglion. These are united with one another by attenuated bands of nerves. Nerve fibres pass also to the ganglion from the cranial nerves and the spinal plexuses. At the coccyx the sympathetic chain terminates in the median ganglion, ganglion impar. From the sympathetic ganglia pass the sympathetic nerves, partly following the blood-vessels, partly continuing farther in the nerves of the cerebrum and spinal cord to the innervated organs (to all smooth muscle fibres). In part they form in the neighborhood of these organs other ganglia (heart-muscle ganglia, ganglia of the mesentery, stomach, etc.).

The arterial blood-vessels of the brain take their origin from the internal carotid arteries and from the vertebral arteries and from the basal circle of Willis, which is formed from the anastomoses of these two (see Plate 5). From it arise the arteries which continue their course in the pia—arteria corpus callosi (for the corpus callosum, median brain surfaces), artery of the fossa of Sylvius (for the environment of the ventricle and basal ganglia), arteria choroidea (for the ventricle), arteria profunda (for the occipital and temporal lobes, etc.). The blood-vessels of the brain substance are what is called terminal arteries, that is, they do not have such profuse connections through anastomoses that neighboring vessels may in case of necessity take their place. The arteries of the spinal cord arise in part from the vertebral artery as anterior and posterior spinal arteries; but in part they proceed from the arteriæ intercostales, etc. The vessels enter from all sides, especially, however, from the anterior longitudinal fissure.

The accompanying illustrations show the divisions of the blood-vessels: Fig. 3 those of the fossa of Sylvius (*A.f.S.*) to the basal ganglia and the internal capsule ($A.l.st.$ = arteria lenticulostriata, which from a pathological point of view is one of the most important branches). Fig. 2, a cross section of the spinal cord with its blood-vessels; one observes the vascular supply of the gray substance through the anterior sulcus artery.

The return of the blood from the brain is accomplished by means of the pial veins which empty in the venous sinuses (from the third ventricle, the vena magna Galeni), which in turn pour into the internal jugular.

In the spinal cord the return circulation is acccom-

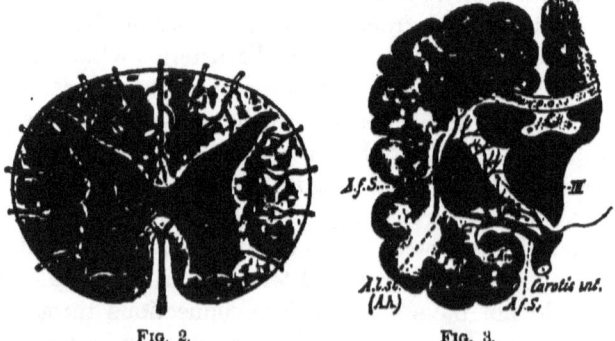

Fig. 2.   Fig. 3.

plished by means of the richly anastomosing Breschet's plexus of veins which ramifies in the dura mater.

The lymph tracts of the brain and spinal cord communicate with the lymph spaces which are distributed throughout the meninges, the ventricular cavities being in open communication with the subarachnoidal space (liquor cerebro-spinalis).

The nerves of the pia and dura are of sympathetic origin, a part of which (those for the dura) arise from the sensory divisions of the trigeminus.

SECTION II.

Development and Structure of the Nervous System.

(Plates 11 to 48.)

THE central nervous system develops from the outer blastoderm, the ectoderm of the embryonic primary constituent. In the middle of the embryonal area a couple of folds develop from the ectoderm in a longitudinal direction. These, coming together and dipping below, form the medullary groove. By the closing of the lateral walls of the primitive tracts, at the dorsal portion, the medullary canal is formed. From the layers of the epithelial cells of the neural canal is developed the matrix or foundation substance (glia) of the central nervous system and the nervous substance proper. The membranes and blood-vessels are of mesodermal origin.

Latero-dorsal from the medullary groove itself there is formed a border studded with cells (ganglion border or furrow), which runs alongside of the medullary channel, throughout its entire length. From the furrow arise the intervertebral ganglia and the analogous ganglia of the sensory cranial nerves, such as the olfactory ganglia, Gasserian ganglion, the jugular, cochlear, petrosal ganglia, etc.

In the first embryonal month the medullary groove divides at its nasal club-shaped end into three primary vesicles. These are the primary anterior vesicle or fore-brain, which later divides into fore- and tween-brain, the primary middle vesicle, and the primary posterior vesicle dividing secondarily into the hind-brain and after-brain. These constitute therefore three, or after later subdivision five, cerebral vesicles. From these develops the entire brain. The cavity of the medullary canal becomes the system of the ventricles, the walls of the canal and vesicles expanding into the structures of the spinal cord and brain.

In the process of development the fore-brain is divided into two parts (the rudimentary cerebral hemispheres) by the falciform process of the dura which grows in from above and presses in the roof of the vesicle (pallium). Through their more rapid growth the hemispheres soon exceed in area the remaining portions of the brain. The wall of the forebrain vesicle expands to form the cortex and medullary substance of the hemispheres, while from its floor the primary constituent of the basal ganglia projects into the primary cavity, which has now become the lateral ventricle. Externally it adheres to the lateral wall (at the external capsule), but mesially it remains free (the nucleus caudatus).

The corpus callosum extends from the vesicles of the hemispheres on their median surfaces, bridges over the middle line, and unites the two hemispheres.

The cortex, which at first is smooth, develops through involution (second to third month) its primary fissures and convolutions, the fossa of Sylvius being the first to develop (see Plate 12).

From the tween-brain arise the optic thalami and the remaining structures such as the hypophysis, conarium, tractus opticus, tractus olfactorius, etc., which surround the cavity that has been transformed into the third ventricle. The thalamus unites at the putamen with the most posterior portion of the basal ganglion. Here, therefore, the fore-brain is in relationship with the tween-brain.

From the mid-brain develop the corpora quadrigemina, tegmentum, and pes of the crura cerebri, while the cavity narrows into the aqueduct.

From the hind-brain develops the cerebellum, and from the after-brain the medulla.

A downward cerebral flexure, the pontine arch, is formed quite early in the mid-brain.

Microscopic investigation of the central nervous system reveals its structure to be of two varieties of tissue: the *matrix*, glia, and the *nervous substance* contained in it, the nerve cells and fibres. Both develop in such wise from the ectodermal epithelial cells of the neural canal that a part of the same, the spongioblasts, increasing from the central canal, send out numerous fibrils in all directions, forming a fine network and thus developing the glia (spider cells, astrocytes).

Cells of different structure, the neuroblasts, develop as germ cells, soon sending out a long, thick process and later numerous finer twigs. From them develop the ganglion cells and nerve fibres. Collections of such germ cells are found in certain definite portions of the central nervous system. Thus they are found in all the cortex of the cerebrum and cerebellum and in the basal ganglia, the latter being looked upon as modified cortex. Further they are located in the thalamus, in the corpora quadrigemina, etc., and finally from the mid-brain in the neighborhood of the central canal as far down as the end of the spinal cord in an arrangement known as the "central nuclear area."

The remaining portion of the central nervous system is developed principally from the long processes of these cells. Collectively these form later the medullary substance, the "white" substance, whereas the aggregation of cells forms the "gray" substance.

The complex "nerve cell with its accompanying nervous prolongations" in its entirety is designated a neuron (τὸ νεῦρον, nerve unit).

The long prolongations of the ganglion cells of the cortex, basal ganglia, and optic thalamus run their course almost entirely within the central organ; that is, they have their beginning and ending there, constituting central neurons. Those from the pericanalicular nuclear areas grow in part from the central organ toward the periphery of the body and constitute peripheral neurons. The last-named comprise the motor cranial nerves, which grow out of the mid- and hind-brain, and the motor spinal nerves which project from the ventral nuclear areas of the spinal cord (anterior horn). The latter develop as the anterior roots from the anterior horn, springing direct, as long processes, from the ganglion cells situated there.

The arrangement of the sensory nerves is exactly reversed. These grow from the cells of the spinal ganglia into the dorsal nuclear area of the spinal cord (posterior horn, etc.) as the sensory spinal nerves.

In a similar way the sensory cranial nerves grow into the brain from the cells of the Gasserian, petro-

DEVELOPMENT AND STRUCTURE. 27

sal, cochlear, and jugular ganglia and from the cells of the retina and mucous membrane of the nose (olfactory bulb, mid-brain, hind-brain). At the same time there pass out from these cells just mentioned other processes to the periphery which constitute the peripheral sensory nerves. This neuron consequently has its cell about in the middle of its course.

In the course of development, the neuroblasts transform themselves little by little into ganglion cells. Their nervous processes may attain to a considerable length. From about the fifth month these nervous processes become surrounded with a sheath which accompanies them throughout their entire course, following the protal formation of the medullary sheaths. Every long process is surrounded with a soft white medullary sheath and becomes thereby a definite, functionating nerve fibre.

Development of the medullary sheath occurs, varying at different times for the different neuron-complexes, from the already named period of embryonal life to the age of childhood and beyond. Successively one portion of the processes after another becomes medullated. Thus results the well-known white color of the medullary substance. As a rule medullation occurs first in those parts that are the earliest to functionate. Thus the first to develop their medullary sheaths are the peripheral reflex tracts, the central sensory conducting pathway before the motor, and the projection fibres before the association tracts, etc. (see Plate 48).

The fully developed neuroglia is made up of numerous small cells, supporting cells, glia cells. The enormous number of prolongations of these cells form a fine fibrillar network in which are embedded the ganglion cells and nerve fibres. The walls of the central canal and of the ventricles are lined with cylindrical epithelium (ependyma), the primitive epithelial cells of this area.

The mature ganglion cells are in part large cells of different shape having well-marked nuclei and made up of protoplasm of very different structure, frequently impregnated with pigment. In part they are small nuclear cells consisting almost entirely of nuclear substance. From the cells there pass off small side prolongations which branch or divide near the cell as do the branches of a tree, and these are called dendritic or protoplasmic prolongations. The process which arises as the direct prolongation of the neuroblast and which has been considered above is known as the axis-cylinder prolongation.

After its exit from the ganglion cell this axis-cylinder prolongation becomes ensheathed with medullary substance and is then known as a nerve fibre.

The nerve fibre consists, therefore, of an axial core or thread, which in turn is composed of numerous individual fibrillæ and the surrounding segmented medullary sheath. These nerve fibres pass out from the central neuron system as peripheral nerves and become covered with a second, very thin envelope around the medullary sheath, known as the sheath of Schwan, or the neurilemma.

The axis-cylinder prolongation may be of various lengths, but it always terminates in an arborization, a splitting up of the fibres. Before this it gives off at different levels lateral branches known as collaterals, which end likewise by splitting up, that is, by terminal ramification. The cells with long prolongations (cells of Deiters' type) make up the preponderating majority, in contrast to those with the very short processes which have their terminal arborization close to the cell itself (Golgi's spindle cells).

Ganglionic cells with their cell branchings, together with their axis-cylinder processes and terminal arborizations, constitute anatomical and physiological units—the neurons.

All nerve tracts are made up of a collection of such neurons. Direct anatomical continuity between neurons does not seem to exist.* They seem to be like the cog-wheels of a machine in which the axis-cylinder termination of one neuron embraces the cell ramifications of the other. The brain, the spinal cord, the peripheral nerves, and the sympathetic nervous system are made up entirely of such neurons.

Transference of excitation from one neuron to another takes place through the occurrence of motion in these terminations. The conduction in the axis-cylinder prolongation is cellulifugal, in the cell ramifications cellulipetal. No neuron is in individual relationship to another neuron only, but to many neurons (intercalary cell neurons), a condition which is made possible by reason of the countless cellular ramifications, and the giving off of numerous collaterals.

The gray substance of the nervous system, the cortex, the ganglia, and nuclear deposits, consist of a groundwork of glia in which are embedded the ganglion cells, their dendrites, and their more or less

---

* In all probability there exists a neuron anastomosis between protoplasmic prolongations.

numerous nerve fibres which are partly medullated and partly non-medullated (white fibres of Remak).

The white substance, the centrum semiovale of the hemispheres, corpus callosum, internal capsules, peduncles, columns of the spinal cord, and peripheral nerves consist merely of medullated nerve fibres, the medullary envelope being responsible for the color and consistence of the substance.

The finer microscopical structure of individual parts of the central nervous system is to be found in the description of Plates 13 to 21 and is far more easy to comprehend from illustrations than from verbal description.

To recapitulate briefly the inner relationships:

In the cell layer of the cerebral cortex the medullary fibres of the centrum semiovale arise and end. They belong to three different systems.

I. THE COMMISSURAL TRACTS.

They constitute the greater part of the central medullary deposit as fibres of the corpus callosum and unite symmetrical localities in the convolutions of one hemisphere with those of the other. Those portions of the cortex that are not thus united through fibres of the corpus callosum, such as the basal, temporal, and occipital lobes, are connected by means of the deeply situated anterior commissure.

II. THE ASSOCIATION TRACTS.

These are made up of medullary tracts which connect different parts of the same hemisphere. Adjacent convolutions are connected by the fibræ propriæ; distant convolutions and lobes by means of the short and long association bundles. They course within the hemispheres from convolution to convolution and are mostly collected into definite bundles, which pass through the white substance to break up in the more distant ones.

Some of these bundles have special names, such as the cingulum, fasciculus uncinatus, fasciculus longitudinalis inferior, etc. The latter unites the temporal lobes with the frontal and occipital lobes, while the superior longitudinal fascicle and subcallosus unite the temporal with the occipital lobes, etc. Portions of the temporal and occipital lobes are connected with especial closeness by numerous association fibres within their own territory and with the remaining portions of the cortex.

III. THE PROJECTION TRACTS.

These are the most important fibre tracts. They arise from all parts of the brain cortex, or end in the same, and pass between the transversely coursing commissural fibres and the longitudinally coursing association fibres further downward. They bring more deeply lying portions of the brain stem and spinal cord into connection with the cortex. The totality of these tracts is called the corona radiata.

The tracts of the corona radiata are divided according to the locality of their axis-cylinder ramifications into short tracts and long tracts.

The corona radiata tracts which come from all parts of the cortex are collected in the internal capsule in the anterior and posterior limbs, and pass further caudad. They thus break through the basal ganglia.

1. The Short Corona Radiata Tracts.

a. One portion of the corona radiata tracts passes from all the convolutions of the cortex into the optic thalamus, coming from out the internal capsule as anterior, posterior, and inferior stalks of the thalamus, and disappears in the thalamus, the corona radiata of the thalamus. A part covers the surface of the thalamus before their entrance, stratum zonale.

b. Another portion passes in a similar way into the nucleus caudatus and there disappears. This portion is to be considered as association tracts.

c. A portion arising from the occipital lobes, the cuneus, collects in the hindermost portion of the posterior limb of the internal capsule and passes to the primary optic centres, to the pulvinar of the thalamus, to the lateral geniculate body, and to the anterior quadrigeminal body. This portion is known as the optic radiation of Gratiolet. From here the optic tract passes farther.

2. The Long Corona Radiata Tracts.

These come from all parts of the cortex, descend in the internal capsule, and pass under the thalamus in the subthalamic region into the tegmentum of the cerebral peduncles, and in part into the crusta.

a. Tegmental Tracts.

A part of these pass directly out of the posterior limb of the internal capsule; another part, after penetrating the inner segment of the lenticular nucleus (ansa lenticularis from the globus pallidus) and breaking through the portion of the internal capsule which lies mesially therefrom in the tegmentum, runs thence as fillet tract farther downward through the medulla as far as its inferior termination, constituting the mesal and lateral fillet tract, central sensory tract.

b. Crusta Tracts.

All of the corona radiata fibres that have not yet branched off pass out of the anterior and posterior limbs of the internal capsule in the crusta and thus out of the hemispheres. The fibre bundles here are considered of three different origins.

1. *The Mesal Bundle of the Peduncle.*—This arises possibly from the cortex of the frontal lobes, passes through the anterior limb of the internal capsule, and ends in the pons ganglia, frontal pontine tract.

2. *The Lateral Bundle of the Peduncle.*—This comes positively from the temporal and occipital lobes, passes through the posterior limb of the internal capsule, and ends likewise in the pons ganglia, the temporo-occipital pontine tract.

3. *The Middle Bundle of the Peduncle.*

THE PYRAMIDAL TRACTS.

These arise from the cortical cells of the central convolutions and their vicinity (posterior parts of the first and second frontal gyri), are collected together in the internal capsule, pass in the knee and anterior third of its posterior limb and then through the crusta, where it lies between the mesal and lateral bundles just described.

This segment is also the only continuation of the crusta fibres which pass through the pons on its posterior border as pyramids and which course farther downward through the spinal cord as the pyramidal tracts. This tract passes in the lateral columns after the larger part of the fibres has crossed over to the opposite side; the uncrossed, smaller portion remains in the anterior columns. This tract is the longest of all the corona radiata fibres, it being possible to fol-

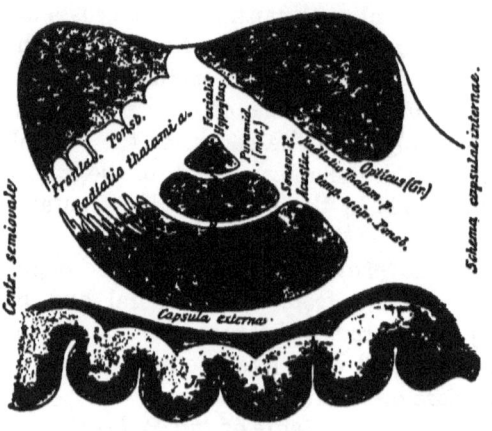

FIG. 4.

low the longest fibres in the lateral columns (crossed) as far as the conus medullaris. From the cerebral peduncles on, this tract continues to give off fibres which pass to the motor nuclei of the opposite side. It constitutes the central motorial tract.

The position of this tract is well illustrated by the accompanying schematic representation of the capsule, which is modified from Edinger (Fig. 4). The direction and position of the tracts are as have been described.

To these corona radiata fibres which arise from the cerebral cortex come yet others that originate in the basal ganglion and optic thalamus, coursing downward as short tracts. To these belong:

1. Those fibres which pass from the nucleus caudatus and putamen to the optic thalamus, to the inner member of the lenticular nucleus, and farther downward to the substantia nigra of the cerebral peduncles.

2. Those tracts which arise from the optic thalamus and pass to the tegmentum and there terminate (red nucleus, substantia reticularis).

In the tegmentum of the cerebral peduncles there begin new fibre tracts. One of these, which passes close beneath the gray matter of the aqueduct, the posterior longitudinal bundle (fasciculus longitudinalis posterior), connecting individual cranial nerves, can be traced into the cervical cord. A larger one likewise passes out from the red nucleus, crosses with its fellow of the opposite side, and goes to make up the brachium conjunctivum (processus cerebelli ad corpus quadrigeminum), passing to the hemispheres of the cerebellum and ending in the corpus dentatum and cerebellar hemispheres.

From the cerebellum fibre tracts pass in or out. Such are the processus cerebelli ad pontem, which crosses to and ends in the pons of the other side, and the processus cerebelli ad medullam oblongatam, the restiform bodies; they consist of fibre tracts from the cerebellum to the olivary bodies and to the posterior and lateral columns of the spinal cord (lateral cerebellar tract), partly long and partly short tracts.

In addition to these the cerebellum, like the cerebrum, contains commissural and association tracts.

The constitution of the spinal cord, as it is made up of the projection fibres of the cerebrum and cerebellum and of the pyramids and restiform bodies, is easily understood from what has been said on page 16. On cross section of the spinal cord we recognize both halves separated anteriorly by the anterior longitudinal fissure. In the median line is the white commissure, passing transversely from one anterior horn to the other; just behind this is the small central canal, posterior to which is the posterior commissure.

The gray substance in the anterior and posterior horns is surrounded by the anterior, lateral, and posterior columns.

1. The anterior column consists of the uncrossed fibres of the pyramidal tract and the anterior ground bundle which is the continuation of a portion of the substantia reticularis of the tegmentum.

2. The lateral column: the crossed pyramidal tracts comprise the lateral cerebellar tracts from the restiform bodies, Gowers' bundle, the anterior lateral bundle, and the lateral limiting layer which comes in part from the tegmentum.

3. The posterior column consists of the column of Goll and the column of Burdach and the root zones.

A more detailed specification of the course and significance of the tracts that have been spoken of, as well as of others not mentioned, follows in Section III. What has been said will be more readily under-

DEVELOPMENT AND STRUCTURE. 37

stood if studied in connection with the serial sections in Plates 24 to 47. The finer relations of the gray substance of the cortex, the basal ganglia, the thalamus, and the nerve nuclei cannot be understood from illustrations alone; plates and text must be studied in conjunction.

In the study of the serial sections it should be remembered, in order to facilitate rapid orientation, that the cerebral cortex is red and the basal ganglia and nuclear deposits are blue.

SECTION III.

Anatomy and Physiology of the More Important Nerve Tracts.

(Plates 49 to 52.)

BY the word tract it is proposed to designate the anatomical substratum in its entire connection, that serves to conduct a physiological excitation from a ganglion cell as far as the excited end organ.

Every individual physiological functionating tract consists of many neuron complexes in connection. Each individual neuron complex has, however, its own anatomical and physiological individuality.

There exist tracts which are made up of two, sometimes three, perhaps more neuron complexes "coupled" one after the other.

The more important of these, the tracts belonging to the projection system of the cerebrum and cerebellum, are the following:

A. TRACTS WITH KNOWN FUNCTION.

1. The motorial, cortico-muscular, centrifugal conducting pathway, consisting of two neuron complexes.

2. The sensory, centripetal conducting pathway, consisting of at least two, probably three or more neuron complexes.

Both tracts come into connection in two parts of their course.

(*a*) There exists an inferior connection: the reflex pathway, independent of the influence of the will, in the subcortical nuclear layer.

(*b*) A superior connection (in the cortex); the tract of conscious will reaction.

The course of the motorial pathway is much simpler and much better known than that of the sensory. We now consider these tracts and their connections in the above-named order:

1. THE MOTORIAL TRACT.

The tract in its entirety is made up of two or more neighboring neuron complexes, the central and the peripheral motor neuron.

The cells of the central neuron, situated in the cortex of the central convolutions, *i.e.*, the pyramidal cells of the cortex, send out their nerve fibres, to the pyramidal tract, in the well-known way through the corona radiata, through the knee and anterior third of the posterior limb of the internal capsule into the crusta. Passing through the pons, they continue as the pyramids.

In the pyramidal crossing, the greater part of the fibres pass into the lateral columns and continue as the tract of the crossed pyramidal lateral column. In the cervical cord this tract lies at the depth of the lateral columns. But it approaches more and more to the periphery in its descent to the conus medullaris, where it terminates. A small portion of this tract passes, without crossing, in the anterior columns and is known as the anterior pyramidal tract. It can be traced only as far as the upper lumbar cord.

Throughout the length of this extended pathway there pass off at all levels between the crusta and the inferior termination definite fibre tracts which cross over to the opposite side (see Fig. 5, which shows the pyramidal tract viewed from below). These fibres ramify and end in the motor nuclei region of the brain stem and spinal cord. The total mass of the pyramidal tract, in consequence of this, diminishes continually as it passes downward. In addition collaterals pass from the continuous fibres to nuclei lying higher up.

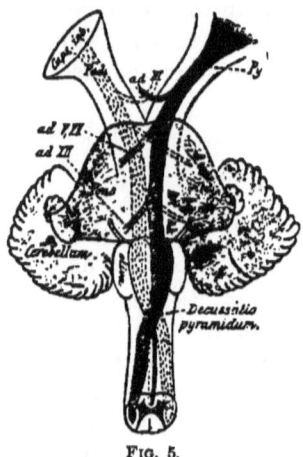

Fig. 5.

The total central neuron mass, therefore, is finally split up within the substations which lie, as has been mentioned, at various levels (segments) of the brain axis and spinal cord. This ramification of the fibres surrounds the cell processes of the ganglion cells of the peripheral motor neuron which begin from this point.

The nervous process of these cells passes out from the brain axis and anterior horn of the same side as the anterior nerve root and runs as a fibre of the motor nerves to the muscle, in the single muscle fibrils of which the terminal ramification of the fibre takes place (motor end plate).

The central motor neuron runs therefore crossed, the peripheral uncrossed. See Fig. 7 of the text. In the tract which is thus made up of these two neurons, the motor impulses that originate in the cells of the cerebral cortex pass to the muscles. In consequence of the decussation of the central neuron, they pass from the cortex of each hemisphere to the musculature of the other side of the body.

Every neuron cell exercises a preservative influence upon its processes, including the long axis-cylinder process. The cessation of this influence results in atrophy of the involved nerve fibre, and even the ganglion cell itself suffers alteration after profound changes within the neuron continuity. The cells of the peripheral neuron exercise in addition a similar influence upon the muscle fibre which it supplies.

In detail, the most important members of the motor tract are the following, stated in order from above downward.

1. *Oculo-Motor Nerve.*—Its central neuron complex arises at some unknown location in the cerebral cortex (gyrus angularis?). It runs through the knee of the internal capsule, separates, however, from the pyramidal tract, in the cerebral peduncles probably as a constituent of Spitzka's bundle (see Plate 35, 2, *s*). The latter arises from the most mesially lying portion of the pyramidal region in the peduncle, turns mesal and upward, and terminates after trespassing beyond the middle line in the nucleus of the motor oculi of the opposite (and of the same?) side, which lies in the central ventricular gray matter beneath the aqueduct of Sylvius. Around the ganglion cells of this locality is found

the terminal ramification of the central neuron of the motor oculi. With these cells and their arborization begins the peripheral neuron. The nerve fibres pass out of the nucleus as the roots of the motor oculi on the same side, in lesser part also decussating with fibres of the other side, and run in the motor oculi trunk to the muscles of the eye. (For further details concerning subdivision and function see explanation of Plate 23.) The position of the motor oculi nucleus, etc., is shown in Plates 18 and 19 and also in Fig. 6 of the text, p. 48. Herein is also to be seen the supposed order of the portions of the nucleus corresponding to the separate eye muscles.

2. *Trochlear Nerve.*—Central neuron runs as in the preceding. After decussation, the terminal ramification of the fibres is found in the crossed trochlear nucleus, below the posterior corpora quadrigemina behind the motor oculi nucleus. The peripheral neuron begins there and passes out as the trochlear nerve behind the corpora quadrigemina, decussating completely with the nerve of the opposite side.

3. *Motor Branch of the Trigeminus.*—Its central neuron originates in an unknown locality of the cerebral cortex (inferior parietal gyrus?), passes through the knee of the capsule behind the nerve fibres of the eye muscles, separates from the crusta in the bundle of Spitzka (?), decussates in the raphe of the tegmentum, and terminates, surrounding the cells of the motor nucleus of the fifth, which is situated laterally in the pontal part of the tegmentum. From here the motor root of the fifth continues its course farther as peripheral neuron. It passes as

anterior root out of the pons and receives also fibres from the motor nucleus of the other side. Even the nasal root of the fifth (descending root) is said to convey motor (trophic?) fibres. This root arises from the lateral portion of the central ventricular gray substance of the aqueduct and of lower parts from large cells situated there until it unites with the anterior root of the fifth as it passes outward.

4. *Abducens Nerve.*—Its central neuron runs as related in 1. Beyond the decussation the terminal endings of the fibre take place in the nucleus of the abducens beneath the floor of the fourth ventricle, near the middle line. The peripheral neuron passes out from its cells as the abducens nerve and runs through the posterior part of the pons.

5. *Facial Nerve.*—Its central neuron complex arises from the cells of the lower third of the central convolutions. The tract passes in the posterior limb of the internal capsule, behind the knee, separates from the pyramidal tract in the bundle of Spitzka (?), decussates in the raphe of the tegmentum, and enters the facial nucleus of the other side which lies laterally, long drawn out in the tegmentum situated deep in the medulla. From here pass upward and inward the single facial fibres as peripheral neuron, and collecting together at the genu facialis they finally bend outward in a great bow and then downward, when they for the first time pass out of the medulla as the facial nerve. This is the lower, mouth-cheek facial.

The central neuron for the upper facial, temporofacial, appears to pass otherwise in some unknown manner. Its peripheral neuron is said to pass out of the hindmost portion of the oculomotor nuclei and to reach the trunk of the facial.

For the orbicularis oris branch the involvement of fibres from the nucleus of the hypoglossus is assumed (?).

6. *Motor Portion of Glosso-Pharyngeal—Vagus.* The central neuron arises from unknown portions of the cortex (middle frontal convolution [?]), passes with the pyramidal tract downward and reaches, after decussating, in some unknown manner the nucleus ambiguus, which it may be assumed is the motor nucleus for this neuron, situated deeply in the posterior portion of the medulla. From this point on the fibres of the peripheral neuron accompany in great part the vagus.

7. *Hypoglossal Nerve.*—The central neuron arises from the lower third of the central convolutions in front of the facial centre, runs downward in the internal capsule behind the tract of the facial, separates in Spitzka's bundle (?) from the pyramidal tract, decussates in the raphe of the tegmentum, and reaches the nucleus of the hypoglossal of the other side, about whose cell it branches. From here the peripheral neuron passes out through the medulla as the hypoglossal nerve.

The tracts described under 5, 6, and 7, particularly those that arise from the left hemisphere, are specially used for the act of speaking. These central neurons, therefore, together comprise the central speech tract; the peripheral, the peripheral speech tract. A particular speech tract has also been assumed, but not demonstrated.

8. *Spinal Accessory.*—The central neuron is said

to come from the basal median frontal convolution, running along with the pyramidal tract; it reaches in an unknown manner the nucleus of the accessorius of the opposite side in the lower portion of the medulla and in the uppermost cervical cord. From here the peripheral neuron passes out as the spinal accessory nerve. An important portion of the nerve, the inner branch, continues farther with the vagus nerve.

9. *The Motorial Tracts of the Upper Extremity.* —Their central neuron complex arises from the cells of the middle third of the central convolutions, especially of the anterior convolution, passes downward through the posterior limb of the internal capsule (anterior third), and continues along with the pyramidal tract through the crusta and pons into the pyramids. In the decussation of the pyramids the greater part passes into the lateral column of the other side. The smaller part (here are found individual variations) passes downward uncrossed in the anterior columns. At the level of the cervical cord the fibres bend in succession at right angles and run out of the lateral pyramidal tracts into the anterior horn of the same side; from the anterior pyramidal tracts through the anterior commissure into the anterior horn of the other side.* Here they ramify about the cells of the anterior horns, especially about the lateral groups and their dendrites.

With these cells the peripheral neuron of these tracts begins. The nerve fibres of these cells pass

---

* This last crossing of the anterior pyramidal tract has been questioned recently, and very properly; in consequence the termination of this tract in the anterior horn of the same side must be regarded as hypothetical.

out as anterior roots and (fourth cervical to first dorsal root) extend to the brachial plexus, from which they pass out, forming the different nerves of the arm (see Plate 23 for explanation). For the position of the cells of the anterior horns and their relation to the single muscles the text figure on page 48 gives the clearest representation.

Above these tracts those for the anterior and posterior neck muscles pass, and under the latter those for the chest and trunk muscles. Their topography can be readily seen by reference to the plate just mentioned. The central course of these tracts is very little known. The centre for the movement of the trunk is in the upper frontal convolutions. For these tracts the descent is not only from the crossed but in all probability likewise from the hemisphere of the same side. Many other muscles, especially those which functionate bilaterally (eye and forehead muscles), receive in all probability innervation from the hemisphere of the same side, and, as has already been described, from the crossed hemisphere. Nevertheless these crossed (or double crossed) central pathways are not in man further known (uncrossed anterior pyramidal tract).

10. *The Motorial Tracts for the Lower Extremity.*—Their central neuron arises in the upper third of the central convolutions and the lobus paracentralis, passes in the pyramidal tract through the posterior limb of the internal capsule (in the middle of the limb), then farther down, as described in 9, undergoes partial decussation, and the greater portion reaches the lumbar region by way of the lateral pyramidal tracts. Here its fibres turn and pass to the

anterior horn, where they have their terminal ramification. From the anterior horn cells, especially from the lateral cell groups, the anterior roots (1–5 lumbar and 1–5 sacral) pass as the peripheral neuron. The latter pass into the lumbar and sacral plexus and subsequently constitute the nerves of the lower extremity; see Plate 23 and text figure 6, in which the relationship of the muscle to the segments of the brain axis and spinal cord is given on the right side.

11. *The Motor Tracts for the Bladder, Rectum, and Sexual Organs* arise from unknown regions in the brain. Their central neuron does not pass downward in the lateral pyramidal tracts, but in all probability in the anterior lateral ground bundle. It ends in the gray substance of the sacral cord. From here the peripheral neuron passes out through the second to fourth sacral roots to the pudendal plexus, etc., and eventually to the muscles of the above-mentioned organs (see Plate 23).

12. To the *sympathetic system* pass central motor fibres, of unknown cortical origin, through the lateral tracts and which appear to terminate in the anterior horns (lateral horns). Here begins the peripheral neuron which goes out through the anterior root, ending in the ganglia of the sympathetic chain and in their analogues. In the cells found here begins the motor sympathetic neuron whose fibres pass to the various unstriped muscles, blood-vessels, intestines, heart (?), glands, etc.

2. THE SENSORY PATHWAY.

As sensory impressions are carried from the periphery to the centre we shall follow that direction

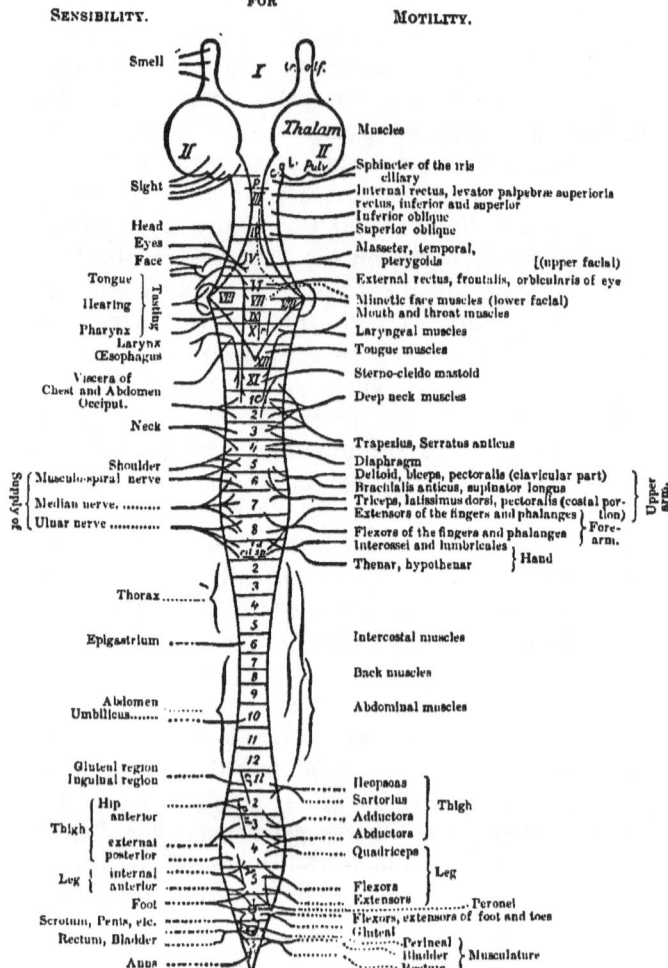

FIG. 6.—EXPLANATION OF THE ABBREVIATIONS.—*tr. olf*=olfactory tract; *c.g.l.*=lateral geniculate body; *p., r., cr., pat., A.,* represent approximately the location of the reflex centres; the pupillary reflex (*p.*), for the respiratory reflex (*r.*), for the cremaster reflex (*cr.*), patellar reflex (*pat.*), and the Achilles reflex (*A.*). The centres for the bladder and rectum in the sacral cord are represented by circles; likewise the centres for erection and ejaculation. The centre for contraction of the uterus is probably likewise in this vicinity.

ANATOMY AND PHYSIOLOGY.

in our consideration of them, starting with the peripheral neuron. The peripheral sensory neuron complex of the extremities and the trunk passes in the sensory fibres of the peripheral nerves (their exact distribution in the skin is shown in text figures 10–12, see also Fig. 6, left side) spinalward through the plexus fibres, and terminates before its entrance into the spinal cord in the cells of the spinal ganglia. From each cell there passes a fibre in the opposite direction. These fibres gather to the posterior roots. The sensory peripheral neuron passes into the spinal cord first by way of the posterior root fibres. They enter in two portions at the border of the posterior horn and posterior column. The two portions are:

1. The small, lateral portion of the posterior root. It enters at the apex of the posterior horn into Lissauer's area. Here its fibres divide into an ascending and a descending branch. Both are short tracts. They pass with a sharp turn into the posterior horn and split up around cells which are there situated.

2. The medial, which is the most developed portion of the posterior root fibres, passes into the lateral portion of the posterior columns (Burdach's column, likewise known as the posterior root zone). Here its fibres fork into ascending and descending branches.

The descending branches are short tracts; they soon turn at right angles into the posterior horn and branching, there terminate. The ascending divisions are in part short, in part long tracts.

The short tracts soon turn and a part of them pass into the posterior horn, while the remainder have their termination in the cells of Clarke's column and the anterior horns. The long tracts pass at first upward in Burdach's columns. Soon after their entrance, however, they approach very much nearer the middle line, principally on account of the recently entering ascending long root fibres. The long fibres that pass into the lumbar cord continue in the cervical cord in Goll's column, to the outer side of which pass the posterior root fibres which enter at a higher level (in the dorsal and cervical cord).

These long, ascending tracts have their terminal ramifications around the cells of the posterior columns of the medulla, the nucleus funiculus Goll, and nucleus funiculus Burdach).

All of these variously coursing fibres constitute together the peripheral sensory neuron complex. The cells of this neuron consequently lie outside of the spinal cord in the intervertebral ganglion. The cell ramification (dendrites) is the very much elongated peripheral nerve, considerably extended in the longitudinal direction, together with its cutaneous terminal branchings. The nervous process is the posterior root fibre with its fibre ramifications that terminate in different parts and levels in the spinal cord.

The fibres split up or ramify about cells that lie:

1. In nuclei of the posterior columns of the medulla (Goll's and Burdach's nucleus).
2. In the posterior horns in their various portions.
3. In the so-called middle zone (between the anterior and posterior horns).
4. In Clarke's columns.
5. In the anterior horns.

From 1–4 inclusive, the central sensory neuron

complex takes its origin. The ramifications that end under 5 may be regarded as reflex tracts. So far all the facts are comparatively certain. Although the further course of the central sensory tract is better understood than formerly, it is still very obscure. The mooted points will be briefly considered later on. The course of the sensory tracts given here will be that which harmonizes best with the postulates of pathology. Anatomically, the course of the sensory tract in man is not yet completely established.

The central sensory neuron begins, in keeping with the terminal ramifications of the peripheral, partly in the already mentioned portions of the medulla, partly in the anterior and posterior horns of the spinal cord. The individual parts will be considered in the order (1–4) cited above.

1. From the cells of the posterior column nuclei, the nucleus of Goll and of Burdach, in the medulla begins the central neuron; the fibres pass as the internal arcuate fibres in form of a bow downward and turn toward the raphe. Here they pass over the median line (fillet decussation the beginning of which is to be seen in Plate 41, 1. It terminates at the level of Plate 39, 2). After this decussation the fibres uniting reach, dorsally, over the pyramids that have already decussated deeper down, to the neighborhood of the anterior ground tracts, where they are known as the layer of the fillet (principal fillet, laqueus superior). The fillet increases in area in its further central course and forms, between the olives, the interolivary layer lying close upon the median line.

Farther up the fillet continues its course in the tegmental region and lies, widely expanded, transversely beneath the same; here it is also known as the median fillet. Thus it continues throughout the pontal tegmentum. In the tegmentum of the crura it swerves completely from the median line and lies lateral to the red nucleus. From this region the superior fillet reaches the subthalamic region, about the level at which the red nucleus terminates. The fillet passes partly through the internal capsule (fibres of the peduncle), running transversely as a part of the ansa lenticularis, breaking through the inner portions of the lenticular nucleus, and rising in the posterior limb of the capsule to the cortex of the posterior central convolutions and parietal lobe. The terminal ramification of the fibres of the neuron takes place about the bodies of cells that lie in these cortical regions. This part apparently is in some manner interrupted in the globus pallidus.

Another portion of the superior fillet emerges apparently from out of the subthalamic region, and passes directly into the internal capsule, in which it ascends behind the pyramidal tracts, terminating in the above-mentioned cortical areas.

A portion or all (?) of these fillet fibres seems to be interrupted in the ganglia of the subthalamic region, the globus pallidus, and the optic thalamus (lateral basal nuclear portion). For cortical conduction it is necessary, therefore, to conceive of a third, centro-cortical neuron (in the optic radiations of the thalamus?). Concerning this, we have no satisfactory knowledge.

This is also called the superior or cortico-thalamus fillet.

2. The central neurons beginning in the cells of

the posterior horns constitute short tracts that terminate for the most part within the spinal cord. Their fibres pass into the border zone of the lateral tracts of the same side (funiculus lateralis of the serial sections, Plate 39, and the following) and also into the posterior tract (ventral field of the same). After a short course upward, in part also downward, they terminate, bending again into the gray substance and splitting up. Similar tracts then connect with these; the conduction, therefore, takes place in the manner of relays in that one short tract connects with another.

In the medulla these fibres enter into the reticulated substance of the tegmentum and terminate in its nuclei.

From the cells of the reticulated substance (nucleus magnocellularis in its entirety) it is supposed that fibres pass to the cortex, annexing themselves to the fillet.

3. From the cells of the middle zone originate three varieties of fibres, partly long, partly short tracts.

a. Fibres that enter into the anterior lateral tract of the same side passing upward in it. One portion is short and terminates bending inward into the gray substance. Another portion passes farther up, reaches the substantia reticularis tegmenti (lying adjacent to the fillet?) fal of the sections.

b. Fibres that enter Gowers' tract of the same side and pass up to the medulla as long tracts. All long tracts send throughout their course at different levels collaterals to the gray substance.

Gowers' tract passes likewise into the substantia reticularis tegmenti of the medulla. Its termination in the brain stem (thalamus?) is still doubtful.

c. Fibres that arise from the median group of cells, especially of the anterior horn, pass (cells of the commissure) to the median line, decussate in the anterior commissure, and pass through the anterior horn of the other side into the anterior lateral tract of this side and pursue their course as long (?) tracts with those mentioned under a. In the medulla they probably join the fillet, no doubt interrupted in the cells of the substantia reticularis of the tegmentum, and run upward with the same toward the cortex.

4. From the cells of Clarke's columns fibres of the central tract pass into the lateral cerebellar tract of the same side (*Cb* in the sections), run upward in the medulla with the restiform body in its most ventral portion and reach with it the medullary subtance of the cerebellar hemisphere, terminating in the vermis (in the crossed nucleus tegmenti?) of the same.

This is therefore in brief the course of the sensory central neurons for the trunk and extremities. We see that a part of the same runs through the fillet direct to the cerebral cortex, crossing in the decussation of the fillet. Another part reaches the cerebellum direct through the lateral cerebellar tract. On the other hand, a third part the anterior lateral tract reaches only to the substantia reticularis tegmenti of the medulla (in part crossed in the anterior commissure); from here on, we must assume a central cortical neuron to the cortex, beginning in the nuclei of the substantia reticularis [?]. These fibres are said to join in part the upper fillet (border of the anterior tract). In the fillet consequently run the chief masses of the central sensory neurons, to the cerebral cortex. The other part passes likewise in the tegmentum (substantia reticularis), at least as far as the thalamus opticus. Further tracts from here are unknown.

Of this description, that portion which gives the details of the decussation of the commissures is doubtful. By many it is denied and only an uncrossed pathway in the anterior lateral tract is accepted. The posterior commissure also contains some decussating fibres, but their significance is likewise not certain. The further course of the fillet beyond the subthalamic region is also doubtful, especially the interrupted portion in the thalamus. The significance of the ansa lenticularis and some other points are hypothetical. Compare with this the accompanying representation of the sensory tracts given in Fig. 7a.

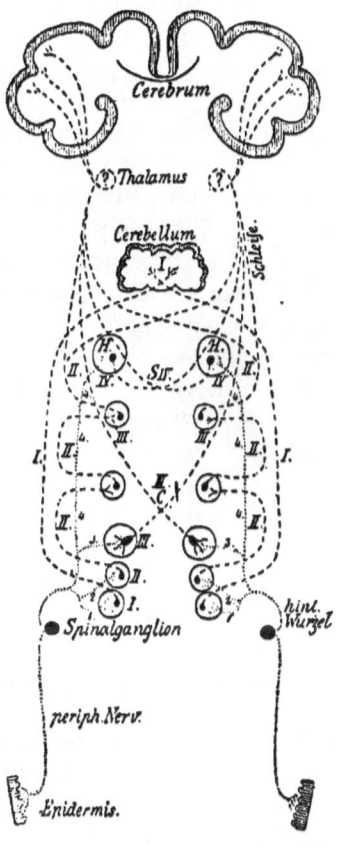

FIG. 7a.

In regard to the conduction of the several sensory

qualities the following is from pathological experience to be posited as probable for man.

The entire conduction of sensibility from the skin, the deeper soft parts in the extremities and trunk, takes place in the peripheral neuron through the sensory nerves, through the cells of the spinal ganglia, by way of the posterior root into the spinal cord. Whether different tracts for the various qualities are already present in this part of the course is not yet determined, but is probable.

We know that sensations of touch go by a different way or at least can go by some other way than do temperature and pain sensations. The last-named are conducted by way of the tracts that enter the posterior horns, etc., and must pass on in the antero-lateral tract (central neuron). A portion of the same certainly passes in crossed tracts (anterior commissure [?]) and reaches with the fillet to the cerebral cortex and thus to consciousness.

Touch sensations and muscle sensations are conducted in the long tracts of the posterior columns to the nuclei of the posterior tracts and thence by the way already described, internal arcuate fibres to the crossed fillet, and from there to cortical consciousness. Conduction in the posterior tract for the touch sense, at least, appears, however, to be replaceable by other tracts (short tracts conducting in the manner of relays). Certain, however, is the further conduction in the fillet (for the most part but not completely in the crossed).

The sensory nerves of the bladder, rectum, etc., pass through the third, fourth, and fifth posterior roots of the sacral nerves into the spinal cord. Their peripheral neurons terminate in the gray substance of the sacral cord. From here (or higher [?] long ascending posterior root fibre [?]) the central conduction begins, ascending probably in the long posterior root tract (Goll's bundle?). As to the further central course nothing certain is known. From the viscera, peritoneum, stomach, mesenteric glands, central fibres pass in the sympathetic plexuses, for example, the splanchnic, into the spinal ganglia and posterior roots and appear to continue their course further in the lateral tract (lateral cerebellar tract?).

Along with the fillet tract of the tegmentum pass also the central neurons for the different sensory cranial nerves, with the exception of olfactory and optic, to the cortex. These are to be considered here in the order of their entrance from below upward.

1. The sensory glosso-pharyngeal-vagus portion. Its peripheral neuron fibres run in the two peripheral nerves to the cells of the petrosal and jugular ganglia, which constitute, therefore, the analogues of the cells of the intervertebral ganglia, and pass out from these in their respective nerve roots through the medulla to the sensory glosso-pharyngeal-vagus nucleus at the posterior end of the fourth ventricle, about the cells of which the terminal fibre ramification takes place. One part does not end at this place but runs as an isolated bundle (fasciculus solitarius [s], descending vagus, etc., root) a short distance downward, and splits up around cells in its neighborhood (posterior horn).

The central neuron originates from these cells and from the cells of the sensory nucleus; it runs toward the raphe, decussates there, and unites with the median superior (according to others the lateral, see below) fillet tract. With this the central neuron complex passes toward the cortex, and splits up about cortical cells at the terminal points of the fillet tract (posterior central convolution [?], for taste fibres in the basal inferior frontal convolution?).

2. Sensory trigeminus. Its peripheral neuron fibres pass out from the three sensory branches of the fifth, unite and pass into the cells of the Gasserian ganglion, and out again as the posterior portion of the root. This neuron passes through the fibres of the pons and ramifies about the cells of the sensory nucleus of the fifth. A part of it, however, passes for a considerable distance as descending caudal (Vc) root of the fifth (formerly falsely called ascending root) into the upper cervical cord, decreasing gradually in size in its course. This descending root of the fifth is seen in all sections of the medulla lateral to the remains of the posterior horn (the substantia gelatinosa of the tegmentum) the uppermost end of which constitutes the sensory root of the fifth. Until the present time the anterior, nasal, root of the fifth, described as motor root, was thought to be sensory; its significance is still uncertain. If, as is very probable, motor, it must also be called descending. Finally another part is said to reach the cerebellum direct from the sensory root of the fifth (direct sensory cerebellar tract). It is therefore to be considered as the analogue of the lateral cerebellar tract of the spinal cord.

The central neuron takes its origin from the cells of the sensory nucleus and from those of the environs of the descending root. Its fibres, decussating in the raphe of the tegmentum, cross the median line and reach the (median ?) tract of the fillet, passing toward the cortex to the posterior central convolution. Here it has its terminal fibre ramification.

3. Auditory nerve. It is divided into two branches, the nervus cochlearis and the nervus vestibularis, each subserving different functions.

a. The cochlear nerve: the true nerve of hearing. Its peripheral neuron begins in Corti's organ in the cochlea, passes into and again out from the cells of the ganglion cochleare lying in the cochlea (analogue of the intervertebral ganglion); as nervus cochlearis it passes to the medulla and ramifies about the cells of the ventral auditory nucleus. Laterally it embraces the restiform body, and continues as central neuron above the corpus trapezoides below the tegmentum. Above the tegmentum the striæ acusticæ pass toward the raphe (doubtful). The neuron decussates and continues its course as the lateral inferior fillet of the corpora quadrigemina, lying external to the median superior tract of the fillet. One part pursues its course uncrossed in the lateral fillet of the same side and comes into conection with the superior olive through fibres of the corpus trapezoides; the lateral fillet passes as far as to below the corpora quadrigemina in the locality already mentioned and ends partly in the posterior corpora quadrigemina and median geniculate bodies (thereto connected with a third central cortical tract?). From here the central acoustic tract reaches the subthalamic region and thence into the posterior limb of the internal capsule, behind the above-described direct sensory superior fillet fibres of the internal capsule. From it the fibres pass into the cortex of the superior temporal lobe.

b. The vestibular nerve. Its peripheral neuron arises from the semicircular canals of the labyrinth where it has its neuron cells and terminates, splitting up about cells of the dorsal auditory nucleus and neighboring cell groups (Deiters' nucleus). One portion constitutes the descending eighth root, while another ascends to the cerebellum as direct sensory cerebellar tract. From the sensory nucleus passes the central neuron (corpus trapezoides?), crossing in the raphe to the median fillet and thus toward the cortex; cortical location unknown.

The optic and olfactory nerves differ from the previously described sensory nerves. They are not to be considered as peripheral nerves but as modified portions of the brain. This is shown by their structure and development.

4. Optic nerve. Its peripheral neuron lies entirely within the external layers of the retina. Adjoining this, its central neuron takes its origin in the internal layers. The neuron fibre runs in the optic nerve toward the centre. A partial decussation takes place in the chiasm in such wise that the fibres from the left retinal halves run in the left optic tract, while those from the right halves run in the right optic tract. The two fibre bundles from the nasal halves of the retina consequently decussate.

Each optic tract contains therefore fibres from both optic nerves, and there are apparently long and short tracts.

The long tract passes in the optic tract to the lateral geniculate body and embracing it runs in the undermost portion of the neighboring posterior limb of the internal capsule direct to the occipital lobes, optic radiation. The existence of these direct fibres is very properly disputed. The short tracts, which are certainly by far the most numerous, undergo in the primary optic centres, corpus geniculatus lateralis, corpora quadrigemina anterior, and pulvinar thalami, terminal ramification about the cells situated there. From these cells the nerve fibres pass out as third (central cortical) neuron, attaching themselves to the optic radiation (posterior limb of the capsule) and go to the cortex of the occipital region, the cuneus. About the cells of this region the fibres of both parts split up.

Other fibres are also found in the optic tract, which arise from the anterior corpora quadrigemina and run centrifugally and for the most part decussate to the retina, the descending optic fibres. Their significance is uncertain.

5. Olfactory nerve. As peripheral neurons are to be considered the nervuli olfactorii, analogue of the intervertebral ganglion, in the special sensory epithelial cells of the olfactory membrane. These run (unmedullated fibres) to the olfactory bulb and split up about cells there situated which compose the glomeruli olfactorii. From here the central neuron passes backward in the olfactory tract. One part runs crossed (commissura anterior), another part uncrossed to the cortex of the gyrus fornicatus and gyrus hippocampus. The fornix, corpora candicantia, and a number of fibre tracts of the tween-brain belong to the olfactory tracts and their association bundles.

This is the course of the sensory tract. Its central neuron runs therefore in greatest part crossed, its peripheral uncrossed (see Fig. 7). In consequence, most sensations reach the opposite hemisphere, but a small part the hemisphere of the same side. This arrangement is entirely analogous to the arrangement of the motor tract.

We turn now to a consideration of the connections between motor and sensory tracts.

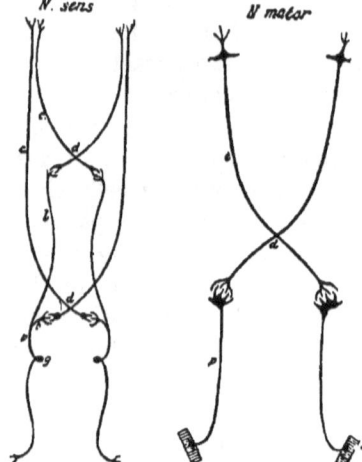

FIG. 7.—*p.* Peripheral, *c.* central neuron: *d*, decussation; *l*, long, *b*, short peripheral sensory spinal tract.

The peripheral motor and sensory neuron (reflex tract) and the central motor and sensory neuron (tract of conscious voluntary reaction) are in physiological connection.

a. The Reflex Tracts.

We understand by reflex, a non-voluntary movement that has been aroused by a sensory stimulus.

The entire process takes place within the peripheral neurons.* These comprise therefore the reflex tracts, reflex arch. This reflex tract is divided into

---

* Whether there are reflex tracts within the sympathetic system (reflex centres in the sympathetic ganglia) is not known.

the sensory portion (sensory peripheral neuron partly) and the motor part (motor peripheral neuron) together with a connecting portion—the reflex collateral—a branch that separates from the sensory neuron after its entrance into the spinal cord or brain stem. This has already been described (p. 50) as a short process of the posterior roots of the spinal cord, passing to the motor cells of the anterior horn (under 5).

The reflex tracts are the first in the fibre primary constituent of the fœtus to become medullated. This corresponds to the fact that the first embryonic processes of movement are of reflex nature.

1. *The Skin and Tendon Reflex Tracts.*—They are in their course the best known. Their sensory limb is formed by the peripheral sensory neuron. Their connecting branch is the reflex collateral that ramifies about the cells of the anterior horns. Apparently not all of the motor cells of the anterior horns serve this purpose but only a part of them. There is to be distinguished a long and a short reflex arc.

(*a*) The short is composed of a collateral that passes direct from the posterior tract, through the posterior horn to the anterior horn (plantar reflex, patellar tendon reflex, spinal reflexes).

(*b*) The long reflex arc arises from the ramifications of a reflex collateral about a cell of the posterior horn. From this cell an ascending and a descending branch with a number of collaterals pass out to one or more motor ganglion cells (in various levels of the anterior horns, and eventually even in the brain stem). In this manner reflex conduction to distant muscle areas is made possible. Hereby is also made clear the possibility of the activity of the brain, that is postulated on clinical grounds for different reflexes. The fibres of the posterior longitudinal bundle appear to functionate in this manner.

The localization of a few reflexes belonging to this class can be seen by referring to Fig. 6, page 48 (for details see Section IV., 3).

2. *The Complicated Reflex Tracts.*—Their detailed course is still very little known.

The deglutition, sneezing, and coughing reflex tracts are composed of the sensory trigeminus-glossopharyngeal-vagus fibres and the corresponding motor nerves of the vagus-accessorius.

The conjunctiva reflex is composed of the trigeminal-facial fibres.

The pupillary reflex is composed of the optic and motor oculi fibres (corpus quadrigeminus and motor oculi nucleus).

The reflex collateral must pass from the sensory nerves concerned to a corresponding motor nucleus. For some reflexes the participation of the cortex is also supposed. Facts are at present wanting for the localization of other important reflexes manifested through the optic, the auditory, etc., nerves.

See further details concerning these reflexes in Section IV., 4.

The functions of the bladder, rectum, etc., under reflex control have their reflex arc in the sacral portion of the cord (see sensory and motor tracts of the same).

b. The Tract of the Will.

Above the reflex arc, which is constituted from the two peripheral neurons, there is as it were a second arc, which is made up of the central motor and sensory neuron together with their connecting pathways in the cerebral cortex. The latter serves for the conduction from a conscious sensation to a voluntarily innervated movement, *i.e.*, for the act of will.

In addition, the central motor neuron exercises an inhibiting influence and the central sensory neuron a controlling influence on the reflex process.

The conscious processes all take place in the cerebral cortex, within which the sensory tracts end and the motor begin. In it also lie the postulated connecting tracts between any two areas of the cortex of different functional significance. A direct ramification of the sensory neuron fibre about the motor ganglion cells is conceivable but is nevertheless improbable. One or more other neurons (transcortical) may very well be interposed between these two. (See plan, Plate 13).

An approximate representation of the nature of the conscious processes in the cerebral cortex is obtained by a consideration of the course of development which the fœtal and infant psychic activity passes through. Although this starts essentially first after birth, yet a series of sensation and reflex processes take place in the fœtal life.

The first movements are produced reflexly. The reflex tracts are medullated early, as has already been mentioned. The entrance of sense impressions from the whole surface of the body into the cerebral cortex, especially that of the posterior central convolution and the parietal lobe, takes place primarily as a foundation for the later activity of the will by way of the tegmental tract (fillet), which becomes medullated long before the central motor tracts. The cells or neurons lying there have the specific property of preserving these impressions in some way through chemical [?] or physical processes, and of reproducing them again under proper circumstances as memory pictures.

In post-fœtal life there come, besides these, memory images of sound waves, from the external physical world through the lateral auditory fillet, that have been stored up in the superior temporal convolutions. In the same way those of visual sensation which are deposited in the occipital lobes, especially in the cortex of the cuneus around the calcarine fissure, coming by way of the optic radiation. A similar condition obtains with smell and taste sensations.

All of these different qualities of memory images that are thus localized in spatially separated portions of the cortex are again connected with one another, by the agency of association processes, with memory images in the same hemisphere, by the association tracts in the strict sense, as well as with others in the other hemisphere, by commissural tracts.

The construction of ideas arises first through the processes of association; every idea is therefore composed of a sum of associated memory images. Associations of a higher order are the ideas whose combination form the processes of logical thought. The reproduction of these associations is called an idea. Only a small part of all possible associations is at any given time in operation; the remainder are latent.

The sum of these associations that are at one particular time operative, plus the sense impressions entering for the first time, make up together the chief component of the content of consciousness. This content is consequently subject to a persistent change. A portion of the associations, supposedly not small, runs its course below the threshold of consciousness (unconscious associations).

In addition to those already named there is still another species of sensation that reaches the cerebral cortex; it is this which is particularly subject to a development into unconscious associations. These are the muscle and articular sensations that are conducted toward the centre at the time, when movement processes take place, which are so early reflexly produced. These store themselves in the central convolutions, probably in the immediate neighborhood of the motor centres, as the sensations of the innervation of those movements that are called forth by the activity of the various muscles, kinæsthetic memory images.

Through the reproduction of memory images of innervation sensations which are being continually deposited in post-fœtal life and through (unconscious) association further developed, the conscious voluntary muscle activity is in some unknown manner made possible and its purposive function controlled. The tract in which this conduction takes place from the terminal portions of the central sensory neuron to those of the central motor neuron (it must lie in the cortex) is the real analogue of the reflex collateral of the peripheral neurons.

An example may illustrate what has been said. On Plate 11, 3, is shown the reflex tract, its sensory part brown; the reflex collateral, its motor part blue.

Above is placed the central tract for conscious activities, its sensory part green, connecting neuron black; its motor portion red. The memory images of sense impressions and muscle sensations store themselves about the sensory (green) cortical ramifications, consequently in the (black) commissural and association neurons of the cortex. These last-named tracts pass to and come from other localities of the cortex.

If the toe be pricked, the leg of the side in the first place draws back reflexly (reflex course in brown neuron to blue, reflex conduction in the lumbar cord), at the same time the sensory stimulation reaches the cerebral cortex in the green neuron. This is associated with various memory images and consequently causes the conscious pain sensation. This causes, now that a new prick threatens association with fresh optical sense impressions, through the arousal of the proper innervation feelings (lobus paracentralis) impulses in the motor tracts (red, blue) that are necessary for the withdrawal of the leg and thereby the muscles that are supplied by it of the different cortical activities that in this wise take place.

The process of vocal speech interests us chiefly here.

In the upper temporal lobe of the left hemisphere are deposited, by the child that is learning to speak, the sound images of heard words (auditory word centre, sensory speech centre). The recognition of the significance of word images, the development of their signification, is first made possible through an extensive association activity.

From here a pathway (connecting part) leads to the **word speech centre, motor speech centre,** in the foot of left inferior frontal convolution Broca's convolution. Here or at least in the immediate vicinity are deposited through practice and repetition the innervation impressions, kinæsthetic memory pictures, for all muscles concerned in the act of speech, the muscles of the tongue, mouth, palate, larynx. The reproduction of these innervation sensations, stimulated from the sensory speech centre, finally arouses the complicated movements that are necessary for the production of letters, words, and sentences. If through some mishap the hearing of a child is lost, it will be impossible for it to learn to talk (deaf-mutism). All of these processes in the acquirement of speech are in close association with the memory pictures from other centres, especially with the optic and the tactile. The more memory pictures that are stored up in the cortex and the more purposive they become through associations, the greater is the intellectual fund of the brain.

The speech centres lie, like the centres for reading and writing which will be considered hereafter, only in the cortex of one hemisphere; in right-handed persons in the left hemisphere; in left-handed in the right hemisphere. The motor speech centre is practically identical with the centre for the facial, hypoglossal, etc.

In learning to read, the scriptoral memory images are stored up in visual centres (occipital lobe) and reach consciousness first through association with the auditory (sound-image) centre and also with the motor speech centres; because in learning to read we always transform the script images into sound images and a speech motor image (reading aloud). In later years many persons may emancipate themselves from these association processes.

In learning to write the scriptoral memory images deposit themselves as sensations of innervation of the writing movement in the right arm centre in the middle third of the central convolution of the left hemisphere. They stand in the closest associative connection with the optic and auditory memory images of this side.

In some individuals the optic, in other the kinæsthetic script associations stand more in the foreground.

The chief regions of the hemispherical cortex in which these processes take place are therefore in the central convolutions, the upper temporal convolutions, the parietal and occipital convolutions. For the higher psychical functions of the intelligence the frontal lobes are designated as the peculiar anatomical substratum. Still one must always bear in mind that the complicated processes of the faculty of thought consist only of more complicated and more extensive association processes, and that the sum of

Fig. 8.

these associations cannot possibly be determined by the activity of a single lobe.

In Fig. 8 can be seen the difference in the distribution of centres for the two hemispheres to the advantages of the left. The latent cortical areas are white. The dotted sense areas contain the psycho-sensory and the psycho-motor areas, the projection tracts along with association tracts. The other white cor-

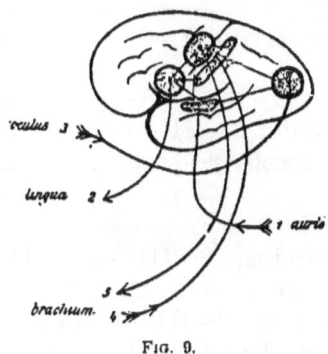

Fig. 9.

tical convolutions contain exclusively association fibres.

For the understanding of speech and other processes Fig. 9 will serve us. The connection of tracts in speaking, reading, and writing is illustrated as follows when we designate: by 1, the sensory speech tract, by I. its centre; by 2 the motor speech tract, by II. its centre; by 3 the optic tracts, III. the optic centre; by 4 the kinæsthetic tract for speech and writing movements, IV. their centres; by 5 the motor writing tract, V. its centre (arm centres).

The construction of ideas is designated by x; this is not supposed to be attached to one centre but is the effect of the united associative activity. A part of the important associations that are in each instance concerned is represented by the designation a in brackets.

Learning to speak: $1\ (a\ 3) - \overbrace{\text{I.} - \text{II.}}^{x}\ (a\ \text{IV.}) - 2$

Learning to read: $3 - \overbrace{\text{III.} - \text{I.}}^{x}\ (a\ \text{IV.}) - \text{II.} - 2$

Learning to write: $3 - \overbrace{\text{III.} - \text{I.}\ (a\ \text{IV.},\ a\ \text{II.})}^{x} - \text{V.} - 5$
Spontaneous speech: $x - \text{II.}\ (a\ \text{IV.},\ a\ \text{I.}) - 2$

Spontaneous reading: $3 - \overbrace{\text{III.} - \text{I.}}^{x}\ (a\ \text{IV.}) - \text{II.} - 2$
Speech repetition *: $1 - \text{I.} - \text{II.}\ (a\ \text{IV.}) - 2$
Writing from copy: $3 - \text{III.} - \text{V.}\ (a\ \text{IV.}) - 5$.
Writing from dictation: $1 - \text{I.}\ (a\ \text{III.}) - \text{V.}\ (a\ \text{IV.}) - 5$.

In order to avoid repetition, other important details will be considered under general symptomatology (Section IV., 3).

It only remains to consider the co-ordination of motor activity.

Every movement, even the smallest, takes place through the combined functioning of a number of muscle groups. The control of this adaptive process of the different muscles involved in every single movement is called co-ordination.

Even in a simple muscular act, agonist and antagonists must act co-ordinately. Much greater co-

---

* If mechanically done then without involvement of (x) associations.

ordination is required in complicated movement processes that are composed of many different single acts following in sequence (for example, walking, talking, etc.). In these cases not only must the single movement be adaptively controlled, but the temporal order must be regulated in proper sequence.

Muscle groups functionating together draw their peripheral as well as their central neuron, or at least a part of these, from cell stations, the individual cells of which are united together in a peculiarly close connection (co-ordination centres and tracts of the brain and spinal cord). Thus, for example, the eye-muscle nuclei and the hypoglossal nuclei are peculiarly close in their connection. The motor impulses from even a single cell may, through collaterals, be conducted to different muscles and thereby support the co-ordinated muscle action.

The control that is exercised by the sensory sensations (particularly muscle sensations) is important for the development of co-ordination. This as well as the influence of the cerebellum upon static co-ordination still remains to be considered.

The cerebellum without doubt exercises an influence upon static co-ordination and equilibration of the body in standing erect and in walking. For this purpose it receives through centripetal conducting tracts the muscle sensations (also optic, tactile, etc.) from the periphery of the body. The restiform bodies conduct the part of the posterior tract which does not reach the fillet from the nuclei of the posterior tract to the cerebellum. In a similar manner the lateral cerebellar tract (function unknown) and fibres from the vestibular nerve (semicircular canals, organ of equilibration?) and trigeminal nerve (direct sensory cerebellar tract) pass to the cerebellum.

But in what manner the cerebellar influence upon co-ordination is to be understood there exist only doubtful hypotheses.

The cerebellum stands in manifold connection with the cerebrum. It is therefore conceivable that it may exercise a controlling influence upon the motor cortical area of the latter and so indirectly on muscular activity. But it is with these centres particularly that the cerebellum appears not to be directly connected.

The fibres of the pons pass from the cerebellum to the crossed ganglion of the pons; in these terminate, as we have already seen, both the frontal and temporo-occipital pontal tracts.

Each cerebellar hemisphere consequently appears to be in direct connection with the frontal, temporal, and occipital lobes of the opposite cerebral hemisphere.

Further, the connecting arm comes from the red nucleus of the tegmentum, which is further connected with the thalamus (fillet [?]), etc., and passes to the crossed hemisphere of the cerebellum. That these cerebro-cerebellar tracts exist we know, but their function is obscure, although they appear to be connected with co-ordination.

A direct influence upon the musculature from the cerebellum is also anatomically conceivable. From the cerebellum, the olivary fibres pass in the restiform body to the contra-lateral olive, and from here the central tegmental tract passes upward and the olivary tracts pass in the lateral pathway downward to the cells of the anterior horns (?).

In conclusion, very little definite knowledge has as yet been acquired in this field of investigation.

SECTION IV.

General Pathology and Therapeutics of the Diseases of the Nervous System.

(Plates 53 to 68.)

1. CAUSATION OF THE DISEASES OF THE NERVOUS SYSTEM.

THE diseases of the nervous system develop either as primary affections of the nerve substance itself, or they follow secondarily to disease of neighboring tissues of the body such as the blood-vessels, skin, bones, etc.

The most common causes of these "secondary" nervous diseases are:

a. Degeneration of the vascular apparatus and its consequences, such as arteriosclerosis, thrombosis, embolism, rupture, and aneurism. As the result of these there follow various forms of nutritive disturbances or mechanical compression and destruction of the nerve substance.

b. Acute or chronic inflammatory processes in the membranes of the brain and spinal cord, such as meningitis, acute, tubercular, or syphilitic; and in the bones of the skull and spine, such as osteomyelitis, caries, and gumma. These conditions may lead up to compression and consequent disturbance of nutrition, and they may also involve the nerve substance directly by continuity.

c. Tumor formation of all sorts, such as sarcoma, carcinoma, osteoma, solitary tubercle, gummatous infiltration, cysticerci, etc., may develop in the involved tissue, the consequences of which in the main are compression of the adjacent nerve substance. Cicatricial formations can injure in a similar manner.

d. Metastatic formations from purulent foci in other organs, such as abscess of the lung, tubercular areas, malignant tumors, pyæmia, etc.

e. Through intoxication after the reception of poisons in other and distant parts of the body and their absorption into the blood, as in diphtheria, septicæmia, erysipelas, typhoid fever, syphilis, nephritis, diabetes, etc.

f. In association with general constitutional diseases, such as anæmia, cachexia, chlorosis, etc.

Those enumerated under *e* and *f* may be considered the most common causes of "primary" diseases of the nervous system. Aside from these there belongs in this category a number of other toxic conditions and infectious diseases which manifest their sinister influence on the nervous system directly; such are lead, arsenic, ergot, alcohol, and acute poliomyelitis, tabes, acute neuritis, etc.

We must likewise consider here all those diseases which are the result of defective protal or embryonal development, such as the muscular atrophies, hereditary system diseases, etc., and the class of diseases known as functional.

We designate as functional those diseases in which we are unable by present methods of investigation to demonstrate the presence of anatomical changes in the nervous system. In these diseases the exact seat as well as the character of the disease are unknown. They have been heretofore considered, correctly or incorrectly, to be disturbance of function dependent upon altered nutrition, molecular or chemical in nature. The scope of these diseases is continually becoming narrower as methods of investigation become more comprehensive. In this category belong hysteria, neurasthenia, and a group of psychical diseases such as melancholia, mania, etc.

In contrast to the diseases designated as functional, diseases in which some anatomical change can be found are known as organic and localizable diseases.

A further subdivision has been made of nervous diseases according to their etiology. They are thus divided into endogenous when their origin is within the human body, and exogenous when the disease develops from external sources, such as the toxic and infectious diseases.

In recognition of the incompleteness of our knowledge of the cause of nervous diseases and especially from didactic and diagnostic considerations, it is urgent to seek for the other causative factors.

For the comprehension of the endogenous diseases, that is, those resulting through defective embryonal development, it is necessary to bear in mind the following: There are a number of diseases which may occur in the various members of a family in the same or in similar forms. In what way the error in the generative series is determined cannot be said.

In senility most organs of the body, such as the heart, kidneys, liver, do not have complete reparation made for the parenchymatous material that is used up by the functionation of the organ, and the result is that an atrophy called "senile" of the organ results. This atrophy is especially marked in the parenchyma of the organ. In a like manner one can imagine that in many individuals whole organs or a part may "wear out" prematurely, especially when such parts have not had an excess of vital resistance from the beginning. In these cases after the reparative ability for the utilized material fails, there comes an atrophy, as in cirrhosis of the liver, atrophic kidney, etc. In various nerve pathways a similar condition goes on. As the result of protal defect of the tract there goes on before the death of the individual, frequently for many years, a slow and uncheckable process which eventually leads to destruction of the tract (involution diseases).

Recently the hypothesis has been put forward that election of any tract for disease or the vulnerability of it is dependent upon its function, and that the tracts that are most active in functionating, such as the reflex tracts, the sensory tracts, and the pyramidal tracts, are the readiest to lend themselves to such degeneration, especially in those who have disturbances of nutrition from other causes, such as cachexia, intoxication, etc.

Aside from these an undefinable inherent or acquired "disposition," which although we do not understand we must admit, exists in the production of the exogenous nervous diseases.

In many nervous diseases some more or less concurrent factors, such as mental strain, excesses of all forms, and before everything else a great number of bodily ailments and diseases, seem to play an important rôle.

2. THE PATHOLOGICO - ANATOMICAL CHANGES IN NERVOUS DISEASES IN GENERAL.

A disease can seemingly without election affect either a certain segment of the nervous system with the nerve fibres and cells of an area which in the brain is called diffuse or focal disease and in the spinal cord transverse disease, or it may involve a number of cells and fibres that have an anatomical or functional connection, or an entire neuron, or indeed an entire pathway, and it is then known as a system disease.

Simultaneous involvement of two or more different neuron complexes or paths constitute the combined system diseases.

In focal diseases an effusion of blood into a certain area and the consequent destruction of nerve substance will cause on the surface of an organ a defect, while in the substance of the tissue there will be a cavity formation in which long after the occurrence the blood stain is recognizable while the effusion is itself resorbed.

From disturbances of nutrition, such as from plugging of the arteries, compression, inflammatory processes, there follows softening, or purulent formations when purulent sources are present. The cells and nerve fibres involved become necrotic and entirely destroyed. The fat and albuminoid corpuscles, the result of this breaking down, are further transformed by the white corpuscles into nucleated cells. Eventually there results a defect, a cyst. Around all such defects the glia tissue becomes thickened into a firm sheath which partly fills up the cyst, and later by means of this process of cicatrization there follows a shrinkage of the focus.

From each such focus of disease there results a series of other disease processes secondarily. There occurs, starting from the point of focal disease, the so-called secondary degeneration in a descending as well as in an ascending direction (different for different pathways), which degenerations are to be followed in part throughout the entire nervous system.

These degenerations develop according to the following principles:

Every neuron cell exercises, as we have already seen, a nutritive influence over its entire prolongation. A destruction of the cell or a disturbance of the relationship of the cell and cell fibre causes a secondary destruction of the distal portion, *i.e.*, the part separated from the cell. The degeneration begins immediately after the disturbance of the connection of the neuron and its prolongation, generally in the peripheral portion of the involved neuron segment but also in the terminal fibre ramifications, and occurs eventually in the entire course of the involved portion. The process of degeneration goes on cellulipetally in those fibres which are separated from their cells. If the solution of continuity be complete and sudden, as from a trauma, the degeneration occurs simultaneously in the entire course of the fibre. If the degenerative process ceases, which after a period from a few weeks to a few months is the case, there results in the spinal cord:

A descending degeneration of the motor (centrifugal) neurons, the neuron cells above the area, and an ascending degeneration of the sensory (centripetal) neurons, the neuron cells beneath the diseased area, and also a degeneration of the tract in the direction of its conductivity.

In addition there exist tracts that may degenerate in both directions—for instance, the fillet tract. This phenomenon has been attributed in part to cessation of function and is known as "functional atrophy." It is possible that these tracts conduct fibres which have different sources of origin. The neuron cell itself suffers secondary changes after the continuity of the fibres has been severed and the performance of its function hindered. For a more detailed consideration of these degenerations see Plate 55, 3, and Plates 61 to 68.

The recognizable microscopical anatomical changes in primary and secondary destruction of the nerve tissues appear to be essentially the same. Aside from the gross morphological alterations, we know generally speaking very little accurately of the more minute pathological process that goes on in the nerve cells and nerve fibres. The ganglion cells suffer a cloudiness of their protoplasm, crumble, shrink, lose their prolongations, and eventually completely disappear.

The nerve fibres likewise become swollen, their medullary sheaths liquefy into drops and eventually become completely resorbed, the axis cylinders dying first. This constitutes the process of degeneration. In the place of the nerve cells and fibres that have disappeared there forms a dense, felt-like formation, newly formed, excessively vascular, glia tissue, the walls of the blood-vessels being very thick. Round-cell infiltration, and in the beginning granular cells are frequently to be seen. Eventually there results in this way the so-called "sclerosis," pigmented connective-tissue formation replacing the former fibres which have suffered disintegration. Instead of the sclerosis there may result simple atrophy of the involved part. This depends in part on the variety of the disease and the rapidity with which the process goes on, and in part on other causes (see description of Plate 64, 2).

3. GENERAL AND SPECIAL REMARKS ON THE SYMPTOMATOLOGY AND LOCALIZATION OF NERVOUS DISEASES. TOPICAL DIAGNOSIS.

The nervous organism can react in two ways to the various disease processes. It can either be stimulated to morbid activity or its function may be encroached upon and eventually entirely annihilated. Frequently both of these procedures go hand-in-hand or rather follow the one on the other, so that in the beginning of the disease the pathological process is of an irritative nature, while a continuance of it causes paralytic manifestations; and symptoms pointing to loss of the function of the organ come prominently into the foreground.

Disease of the motorial cortico-muscular tract can call forth as irritative symptoms the severest convulsive attacks, convulsions, tonic or clonic, fascicular or fibrillary muscular twitching; and as paralytic symptoms, weakness, paresis, or paralysis of single muscles or of members of the body, depending upon the extent of the disease.

Disease of the sensory tract can have its irritation

symptoms manifested by neuralgic pain, pathologically increased sensitiveness, such as hyperæsthesia and hyperalgesia, and its paralytic symptoms by weakness or entire destruction of sensibility or a qualitative destruction, such as anæsthesia, analgesia, and thermo-anæsthesia.

Disease of the pathways of co-ordination may be manifested by the limitation or cessation of co-ordination of muscular action, causing ataxia, ataxia of movement, and static ataxia.

The variety of the disease process is of less value in determining the beginning symptoms than the locality and extent of the disease. And here it is important to differentiate between focal and system diseases; the symptoms of the latter are very much more constant and well-defined than those of the first, which vary according to the topography of the focus.

The fundamental facts necessary for a comprehension of the symptomatology of nervous diseases are the following:

A lesion of the central motorial neurons is followed by a paralysis of the muscles which they supply and voluntary muscle response perishes. The paralysis, in accordance with the crossing of the central neurons, must be manifest on the opposite side of the body. The paralysis is of a spastic nature; that is, the muscles affected resist passive movements, the tonus of the muscles is accentuated (hypertonia), and the muscles have a tendency to spontaneous shortening, the so-called contracture formation. Although there follows an enormous inactivity atrophy, there is no degenerative muscle atrophy, that is, the relative protoplasmic make-up of the muscle remains unaltered.

The reflex tracts are not interrupted, the occurrence of the reflexes are facilitated, and increased reflexes are the consequence. If the seat of the disease be in the vicinity of the cell bodies of motorial neurons (cortex of the brain, central convolutions), the resulting irritative symptoms may be manifested as involuntary contractions of the muscles to which they are distributed—that is, motorial irritative manifestations, tonic and clonic convulsions, epileptiform convulsive attacks; half-sided or bilateral, slow, involuntary, choreic, irregular movements; spontaneous, determinate, athetoid movements of the fingers, etc.

A lesion of the peripheral motorial neuron leads likewise to a paralysis of the muscles to which these neurons are distributed. The paralysis, however, in accordance with the uncrossed course of the neurons, is on the same side of the body as the lesion. The paralysis is of a flaccid nature, that is, there is no resistance to passive movement, the tonus of the musculature is lost, and there is no tendency to contracture. The muscle undergoes, in a longer or shorter time after the onset of the disease, depending upon the nature of the lesion, a degenerative atrophy. The protoplasm of the muscle is involved, it becomes granular and is resorbed, the muscle changes into a fibrous mass and the protoplasm of the muscle fibre completely disappears. See also under electro-diagnosis, part 4 of this section.

If the reflex arc is interrupted the reflexes are lost. As irritative symptoms there often appear fibrillary muscular contractions in smaller muscular bundles, which occur without producing any movement of the part, and as a result of this the muscle may pass into a condition of fluctuation, of unrest, which, however, eventually disappears by itself.

Lesion of the sensory tract entails disturbances of sensibility on the opposite side of the body in disease of the central neurons and on the same side in disease of the peripheral neurons. Aside from this there is no essential difference in the resulting symptoms, such as there is, for instance, in disease of the motor tract. Attacks of severe, neuralgic, lancinating pains are considered somewhat significant of involvement of the posterior roots (peripheral neuron.)*

The motor pathway runs through the entire nervous system and in every point of its course it contains central and peripheral fibres, the latter going successively at all levels toward the periphery. Therefore we use, in making exact localizing diagnoses of focal areas of disease, the above discussed motorial symptoms in disturbance of the central tract in much the same manner as the axis of abscissas on which other focal symptoms indicate the height of the lesion as ordinates.

We shall consider only the focal and system diseases.

I. Symptoms of Focal Disease.

A. OF THE BRAIN.

These are divided into the direct focal symptoms in the narrowest sense, and into the indirect, which are operative remotely as well as in neighboring portions of the brain, and into general symptoms, such

---

* Pain of central origin may occur and be of great severity.

as fever, cachexia, vomiting, headache, disturbance of physical integrity, depending upon the form and the development of the lesion. In very extensive lesions there follows loss of consciousness, coma, or somnolency. In the following only the direct symptoms naturally are considered.

1. Symptoms in Lesion of the Cerebral Cortex.

a. Of the Frontal Lobe.

Disease of this part of the brain occurs frequently without giving rise to symptoms. In foci of considerable extent there is oftentimes psychical disturbance, such as apathy and dementia.

If the inferior frontal lobe of the left hemisphere in its posterior portion is involved there will be accompanying loss of voluntary speech innervation, constituting motor aphasia.

With a more extensive lesion, if the arm centre in the central convolution is involved, there may be also inability to write voluntarily, constituting agraphia.

With a lesion of both frontal lobes there will be disturbances of bodily movements and disturbance of phonation.

b. The Central Convolutions.

Disease of the upper third of the central convolutions and of the paracentral lobe is followed by spastic paralysis of the entire lower extremity of the opposite side, crural monoplegia, and eventually such irritation symptoms as clonic convulsions and choreic, athetoid movements of the leg.

Lesion of the middle third results in crossed spastic brachial monoplegia and later irritation symptoms in the same arm.

Lesion of the inferior third produces crossed facial monoplegia and eventually twitchings in the parts supplied by the seventh nerve. Only the inferior portion of the musculature of the face is involved.

Lesion in the region beneath and in front of the facial centre produces lingual monoplegia, paralysis of the opposite hypoglossal; the tongue when protruded turns toward the opposite side.

When the extent of the focus in the central convolutions is a larger one and involves more than one centre, then of course more groups of muscles of the opposite side are affected, causing partial hemiplegia, total hemiplegia, and a complete lesion of the extremities, of the face, and of the tongue. The eventual irritation symptoms that may occur from cortical lesions in this area are partial cortical epilepsy (Jacksonian epilepsy), the disease causing from the primary focus a successive irritation of all the motor centres in the order of their situation. Accordingly the spasms take place on the opposite side of the body. The irritation may be transferred to the other hemisphere (through the commissural fibres). When there are focal lesions in the cortex of the central convolutions of both hemispheres, then a double-sided spastic paralysis takes place, facial, brachial, diplegia, etc. (see pseudo-bulbar paralysis).

The muscular paralysis in these cases is a spastic one. The tendon reflexes of the paralyzed side and often of the other side are increased, the skin reflexes of the paralzyed side are often diminished.

c. The Parietal Lobes.

Lesions of these parts are often accompanied by disturbance of sensibility in the opposite half of the body, especially disturbance of the muscle sense and of cutaneous sensibility. Nevertheless the nature of this disturbance (hemianæsthesia) is not sufficiently elucidated. A focus in the angular gyrus can produce disturbance of the ocular muscles of the opposite eye, ptosis, and equinanimous deviation of both eyeballs toward the seat of the focus (conjugate deviation).

Lesion of the left inferior temporal lobe, the supramarginal gyrus, may cause inability to read (alexia), while the ability to speak is in nowise disturbed.

d. The Occipital Lobes.

Unilateral destruction is followed by the disturbance of vision known as hemianopsia, *i.e.*, blindness of similar sides of the retinæ. For instance, destruction of the left occipital lobe causes blindness of the left half of both retinæ, or, what is the same thing, loss of vision in the right half of the two visual fields, thus constituting right-side hemianopsia. In multiple peripheral lesion of the left hemisphere there develops loss of the optic memory picture and the capability of their association, thus causing psychical blindness, inability to grasp the conception of a seen thing or object, and eventually even to recognize it, *i.e.*, optic aphasia.

When the focus of disease is on both sides there may result complete (cortical) blindness.

e. The Temporal Lobes.

Destruction of the superior convolution of the left hemisphere is accompanied by loss of the power of recognition of word sounds and their constructive association, thus constituting sensory aphasia (loss of the understanding of words with preservation of the power of speech).

With bilateral lesions there occurs "cortical deafness."

f. Mesal Hemisphere Surface.

The paracentral lobe is to be reckoned with the central convolutions, the cuneus with the occipital lobes.

Destruction of the uncinate gyrus (gyrus hippocampus) causes disturbance of the sense of smell, central anosmia. Lesion of the posterior basal portion of the frontal lobes results in disturbance of the sense of taste (central ageusia [?]), which, however, is apparent only when the focus is on both sides.

g. Island of Reil.

Lesion of the left island of Reil produces a disturbance in the ability to correlate the speech centres, so that the result is defective word and sentence formation, lapsus linguæ, etc. (paraphasia). It destroys the control of the sensory speech centres.

2. LESIONS OF THE CENTRUM SEMIOVALE.

These lesions are often symptomless. When tracts which go to the cortex are affected just before passing into the latter (subcortical lesions) the same symptoms then may occur as when the cortex itself is involved (but as a rule without any symptoms of irritation), such as monoplegia, aphasia, hemianopsia (optic radiations), and hemianæsthesia. As the motorial pathways converge more and more as they pass in their course through the white substance, a small lesion will cause more extensive symptoms the more deeply it is situated. For instance, while a cortical lesion of a certain size may cause only monoplegia, the same lesion may cause, if deeply seated, complete hemiplegia.

Lesions of the corpus callosum are generally symptomless.

3. Lesion of the Internal Capsule.

a. Destruction of the anterior limb occurs without attributable symptoms.

b. When the lesion is in the posterior limb and only the knee is involved, the symptoms resulting are: Paralysis of the facial (inferior branch), paralysis of the hypoglossal of the opposite side. (When the lesion is on the left side there is disturbance of speech.) When the lesion is double-sided there is diplegia. When the middle portion is involved there is crossed hemiplegia of the brachio-crural type, and when the posterior portion is the seat of the lesion there will be partial hemianæsthesia of the opposite side of the body and often hemianopsia, but the latter is by no means certain. There may likewise be lessened acuity of hearing. In most instances the focus is so extensive that the larger portion of the segment of the capsule is destroyed; then there occur the well-known symptoms, total hemiplegia, facio-hypoglossal-brachio-crural type, of the opposite half of the body with hemianæsthesia.

The ocular muscles, muscles of mastication, and those of the upper face, throat, and trunk, muscles that functionate bilaterally, are involved only if the lesion be a double-sided focus.

4. Lesion of the BASAL GANGLIA, nucleus caudatus and nucleus lentiformis, may apparently be symptomless; the symptoms which so often accompany lesion of these parts being attributable to involvement of the adjoining internal capsule and may, therefore, be called indirect symptoms.

5. Sequentially to LESION of the THALAMUS, especially if the focus be double-sided, there follows disturbance of psychical functions, especially of the psychical reflexes, such as laughing and crying.

There are also disturbances of co-ordination manifested by choreic movements, not infrequently rapidly developing muscular atrophy on the opposite side of the body.

Destruction of the posterior portion of the thalamus, the pulvinar, causes partial hemianopsia.

6. LESION in the SUBTHALAMIC REGION causes crossed hemianæsthesia (lesion of the fillet).

7. LESION OF THE CORPORA QUADRIGEMINA.

As an irritation symptom there appears a peculiar disturbance of locomotion, a reeling, uncertain gait, a cerebellar ataxia, which is possibly explained by involvement of the cerebellar peduncle or by influence on the neighboring cerebellum. Such a gait in combination with other symptoms speaks in favor of lesion of the corpora quadrigemina. These symptoms are paralysis of the ocular muscles, ophthalmoplegia. The latter can be the result of nuclear involvement or of involvement of the nerve fibres, and the seat of the lesion may be unilateral or bilateral. Lesion of the most anteriorly situated cells of the nucleus would seem to cause paralysis of the interior eye muscles, the ciliary and sphincter pupillæ muscles, while lesion of the anterior inferior portion of the corpora quadrigemina paralyzes the internal and superior rectus, the levator palpebræ superioris; the remaining eye muscle nuclei being situated more posteriorly.

If the tegmentum beneath the corpora quadrigemina is simultaneously involved there results crossed (incomplete) hemianæsthesia (fillet).

If the crusta are damaged there results crossed hemiplegia of the brachio-crural type and also of the facio-lingual.

If the lesion is only one-sided, there results a total crossed hemiplegia with paralysis of the motor oculi of the same side (peripheral same-sided neuron). This form of hemiplegia which involves the crossed descending central neuron for the hemiplegia as well as the uncrossed peripheral neuron for the oculomotor paralysis is known as hemiplegia alternans. Where the disease focus is double-sided, other symptoms bespeak the extensiveness of the lesion.

The eye muscles that are thus paralyzed undergo a degenerative atrophy, while the paralyzed extremities develop a spasticity without severe injury to the muscles.

If the optic tract or the geniculate body be in any way involved, the result will be hemianopsia.

In destruction of the posterior corpora quadrigemina and median geniculate bodies, there develops frequently disturbance of hearing.

8. Lesions of the Pons.

One-sided focal lesion can cause hemianæsthesia (fillet) and hemiplegia of the brachio-crural type (pyramidal tract) of the opposite side of the body.

Further, paralysis of the masticatory muscles on the same side (motor trigeminus, peripheral neuron) may occur or when more posteriorly the muscles innervated by the facial nerve of the same side (peripheral neuron) may be affected, thus constituting hemiplegia alternans.

Therefore anæsthesia in the trigeminal distribution of the same side, disturbance of the sense of taste, articulatory speech disturbance (defective letter formation) may result from involvement of the facial-hypoglossal fibres.

As irritative symptoms there may be vertigo, ataxia, and trismus.

When the focus is double-sided there will be corresponding symptoms.

9. Lesions of the Medulla.

When the lesion is one-sided there may result hemiplegia of the brachio-crural type and hemianæsthesia of the other side of the body. If the internal arcuate fibres in the posterior segment of the medulla are involved there may be also hemianæsthesia of the same side, by involvement of the crossed fillet fibres.

Aside from these there may be paralysis of the hypoglossal of the same side (peripheral neuron, hemiplegia alternans), the tongue when protruded deviating toward the side where the focus is situated. The involved half of the musculature of the tongue undergoes degeneration.

Consequentially to this there is articulatory speech disturbance, bulbar speech (without the existence of central speech disturbance) which is known as dysarthria labio-lingualis.

Furthermore, there may result paralysis of swallowing, disturbances of respiration, aphonia (paralysis of the vocal cords), and disturbances of the circulation, especially if the foci are bilateral. Destruction of the inferior olive is followed by difficulty of equilibration.

10. Lesions of the Cerebellum.

The symptoms are often very dubious and unclear.

Disease of the vermis is followed by disturbance of equilibrium, attacks of vertigo, cerebellar gait (reeling gait), vomiting, etc.

Lesion of the processus cerebelli ad pontem is also followed by failure to maintain equilibrium, by vertigo, compulsory movements such as turning on the long axis, etc.

Lesions of the hemispheres may occur without symptoms.

11. Focal Diseases of the Base of the Brain.

These are characterized especially by the simultaneous development of paralytic symptoms in several of the basal cranial nerves on the side corresponding to the lesion, or on both sides, such as the motor oculi, the abducens, facial, trigeminus, hypoglossal, etc.

At the same time the pes pedunculi cerebri, the tractus opticus, and the chiasm may also be involved.

The symptoms will therefore be very variable and are easy to reconstruct. See further under peripheral nerves of this section. Most frequently the nerves that are simultaneously involved are the second and third, the sixth and seventh, and the tenth, twelfth, eleventh (anterior, middle, and posterior fossa of skull).

To recapitulate briefly: the typical brain paralysis is spastic hemiplegia and hemianæsthesia of the opposite side).

Cortical foci cause preponderatingly monoplegias; capsular foci, hemiplegias pure and simple; foci in the corpora quadrigemina and in the cerebral peduncles cause hemiplegia with crossed motor-oculi paralysis (hemiplegia alternans superior).

Foci in the pons: hemiplegia with crossed trigemino-facial paralysis (hemiplegia alternans media).

Foci in the medulla: hemiplegia with crossed hypoglossal, etc., paralysis (hemiplegia alternans inferior).

All of these indicate unilateral lesion.

As the projection fibres in the brain roots come closer and closer together, so it comes that the lower the lesion the greater the possibility for a unilateral lesion to produce double-sided symptoms; the truth of this is more striking for foci in the spinal cord.

B. FOCAL SYMPTOMS OF THE SPINAL CORD.

In contrast to the typical cerebral palsy, hemiplegia, the typical palsy of spinal origin is paraplegia, *i.e.*, paralysis not only of one extremity but of both simultaneously.

This results because of the fact that the motorial pathways for both sides of the body pass in their course close to one another and it becomes easy for one focus to involve both.

Nevertheless there occur half-sided spinal-cord injuries, from stab wounds, tumor formation, etc., and these have, to differentiate them from the ordinary spinal-cord disease, a different but very typical symptom complex known as Brown-Séquard's paralysis.

As the result of a unilateral spinal-cord lesion there develops a paralysis of the extremity of the same side, varying according to the seat of the lesion, while the sensory disturbances are on the opposite half of the body; this because of the fact that the motor fibres in the spinal cord are in a great measure decussated (crossed pyramidal tract), but the central sensory pathway in a great measure is not (antero-lateral bundles, anterior commissure, etc.). Aside from these there results a small anæsthetic zone at the level of the lesion and on the same side, the result of injury to the uncrossed peripheral sensory neurons of this side, and likewise occasionally a small girdle-shaped area of hyperæsthesia which must be looked upon as a symptom of irritation. Frequently this symptom of unilateral lesion of the spinal cord is not of such constant occurrence as the above mentioned.

We shall consider now the more prominent and common focal lesions.

a. Lesions of the Cervical Cord.

These cause, when the entire cross section of the cord is involved, a complete paralysis of both arms and legs, universal paraplegia in connection with complete anæsthesia of the trunk and extremities. The paralysis of the legs is spastic in nature, the result of involvement of the central neuron; while, on the other hand, the paralysis of the arms, if the seat of the lesion be from the fifth cervical to the first dorsal, is of a flaccid nature, due to the involvement of the peripheral neuron of the anterior horn. Those muscles of the upper extremities whose cell nuclei in the anterior horns are destroyed undergo degenerative atrophy. The reflexes of the upper extremities are lost, owing to interruption of the reflex arc, while those of the lower extremities, on the other hand, are increased, due to the loss of the central inhibiting fibres.

Sensibility is destroyed on both sides as far up as the level of the entrance of the posterior roots above the diseased segment.

Aside from these there is disturbance of the bladder and of the rectum, incontinence of urine and feces, the result of interruption of the central conducting pathways.

The intensity of the symptoms varies according to the extent of the lesion.

The flaccid paralysis of the upper extremities involves different or all the groups of muscles, depending upon the seat of the focus of disease. This is illustrated by text figure 6, Section III., p. 48. The most important segments of the cervical cord that become involved are:

α. Fourth segment: paralysis of the diaphragm.

β. Fifth and sixth segments: paralysis of the upper-arm type, deltoid, biceps, brachialis anticus, supinator longus, and eventually paralysis of the shoulder muscles.

γ. Seventh and eighth segments: paralysis of the lower-arm type, the muscles supplied by the ulnar and the triceps.

δ. Eighth cervical and first dorsal segments: paralysis of the small muscles of the hand and eventually oculo-pupillary symptoms (see under C., 1).

In all these muscles degenerative atrophy takes place.

b. Lesions of the Dorsal Cord.

In these lesions the upper extremities are not affected except when the first dorsal segment is the seat of disease, the small muscles of the hand are involved.

There is a spastic paralysis of the lower extremities, the reflexes are increased (patellar clonus, ankle phenomenon); from the level of the body downward from the place where sensory fibres enter above the focus of disease there is complete anæsthesia, analgesia, thermo-tactile anæsthesia, diminished muscle sensibility, etc.

In addition there is paralysis of the bladder and rectum. Those muscles of the trunk which are cut off from their spinal-cord innervation by the focus of disease undergo a degenerative atrophy; just what muscles it is difficult to demonstrate (intercostal, lumbar, dorsal, abdominal muscles, etc. See under electro-diagnosis, p. 116). When such paraplegia is of long standing the tendon reflexes may become very much limited owing to secondary contractures in the extremities (?).

c. Lesions of the Lumbar Cord.

These cause disturbance of mobility and sensibility of the lower extremities only.

This consists of a flaccid paralysis (involvement of the peripheral neuron) with complete anæsthesia.

The muscles of the legs undergo a degenerative atrophy, the reflexes are lost.

There is total paralysis of the functions of the bladder and rectum (preservation of the sphincter reflex, see page 133).

If the focus of disease is in the upper lumbar cord there will be degeneration of the muscles supplied by the crural nerve, the quadriceps and psoas, as well as of those supplied by the sciatic, the gluteal, peroneal, and calf muscles. For further elucidation of these muscle paralyses see text figure 6, p. 48.

d. Lesions of the Sacral Cord.

With these lesions the thighs remain intact, while some of the small foot muscles become paralyzed and atrophied. There is anæsthesia of the outside of the feet and of the toes, and likewise of the anal region. There is complete paralysis of the functions of the bladder and the rectum owing to loss of the sphincter reflexes. The patellar reflex is preserved, the focus being too low down to involve the reflex arc.

e. *Lesion of the Cauda Equina.*

The symptoms are, in short, those of involvement of the lumbar-sacral cord, as the cauda is really a collection of nerve fibres that spring from this part of the cord, having merely a lower location. If the upper fibres that pass to the thigh are spared, there will be essentially a sciatic paralysis, that is, flaccid paralysis of the leg and peroneal muscles, also of the flexors of the thigh and eventually of the gluteal muscles and of the small muscles of the feet. In addition there will be disturbances of sensibility in the areas supplied by the sacro-sciatic nerves.

Total paralysis of the bladder and the rectum occurs in many cases of disease of the cauda. The patellar reflex is preserved, the Achilles tendon and sphincter reflexes being lost.

If with this disease of the spinal cord the posterior roots corresponding to the diseased segment are involved, there will be in addition severe pain of a neuralgic character in the peripheral distribution of the same, but this frequently bespeaks simply an involvement of some roots above the area of the focus. Frequently there appear reflexly produced involuntary muscular twitchings in the paralyzed extremity.

C. SYMPTOMS IN LESION OF THE PERIPHERAL NERVES.

Lesion of the peripheral nerves is followed by motor and sensory symptoms which bespeak exactly the extent of involvement by the lesion, and the recognition of this is therefore indispensable for an exact diagnosis (see Plate 23).

The paralysis resulting from lesion of these structures is a flaccid one; the muscles involved undergo atrophy of degeneration when the process is at all severe. The sensory disturbances that result are limited exactly to the distribution of the cutaneous nerves, and frequently there is excessive sensibility, painful in character, as well as abnormally acute sensibility-paræsthesia (formication, sticking, feeling of fur, burning, etc.). The reflexes are lost or diminished if their tracts pass through the involved nerves.

I. Plexus Paralysis.

The symptoms of plexus paralysis are a mixture of the paralytic phenomena of the nerves collectively involved. The consequences are that the symptoms vary a great deal; nevertheless when the lesion is in certain locations they are sufficiently characteristic to be designated by special names. They are:

a. Erb's type of brachial palsy. In this the involvement is of the fifth and sixth cervical roots, destruction of which causes a paralysis and atrophy of the deltoid muscle, biceps, brachialis, supinator longus, supra- and infraspinatus. Thus results inability to raise and abduct the arm and to flex the forearm.

b. Inferior plexus paralysis of the brachial plexus results when the eighth cervical and first dorsal roots are the seat of disease. It causes a paralysis and atrophy of the small muscles of the hand and anæsthesia in the distribution of the ulnar nerve. In addition to these manifestations there may be oculomotor pupillary symptoms (first dorsal segment), such as myosis, narrowing of the palpebral fissure, sinking in of the eyeball. The lower extremities are not involved, in contrast to the symptoms which result when the corresponding segments of the cord are diseased.

II. Lesions of the Cranial Nerves.

The area to which the sensory manifestations of the different nerve trunks are limited is shown in text figures 10 to 12. In detail their lesions call forth the following symptoms:

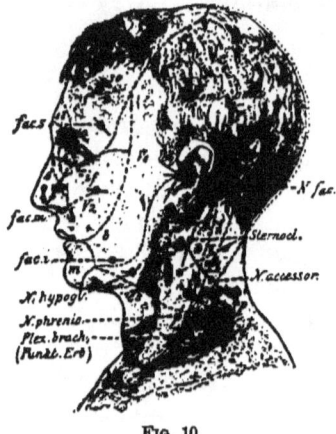

Fig. 10.

1. *The olfactory nerve:* anosmia on the corresponding side of the nose, also parosmia (disagreeable odors).

2. *The optic nerve:* amaurosis of the corresponding eye, of the entire visual field; eventually only lack of visual acuity; narrowing of the visual field; atrophy of the optic nerve (white atrophy); mydriasis; reflex pupillary sluggishness; consensual limitation. This in contrast to the condition of hemianopsia and half-sided blindness which results when the optic tracts are diseased.

Lesion of the chiasm causes bitemporal hemianopsia, due to destruction of the inner retinal bundles which cross here.

3. *The oculomotor nerve:* ptosis, paralysis of the

levator palpebræ superioris, immobility of the eye in an inward and upward direction (internal, superior, and inferior recti, and the inferior oblique), crossed diplopia, fixation of the eyeball on turning it outward on account of contracture of the fixed rectus internus; immobile pupils, and mydriasis the result of interference with accommodation and the sphincter action through ciliary muscle paralysis.

4. *Trochlear nerve:* diplopia when the eyes are turned downward (superior oblique muscle).

5. *Trigeminal nerve:*

a. Supraorbital branch: anæsthesia of the skin of the forehead, of the conjunctiva, and back of the nose (see Fig. 10, V_1).

b. Infraorbital branch: anæsthesia of the cheeks, of the alæ of the nose, and of the gums, and disturbance of the sense of taste (Fig. 10, V_2).

c. Inframaxillary branch: anæsthesia of the skin of the face over the lower jaw, of the tongue, and the mucous membrane of the mouth (Fig. 10, V_3). Also disturbance of the sense of taste (anterior segment of the tongue), paralysis of the muscles of mastication, disturbance, in the secretion of saliva, trismus, *i.e.*, cramp condition of the masticatory muscles, an irritation symptom.

The irritation symptom of the sensory branch is neuralgia of the trigeminus, which may be of extensive distribution, depending upon the seat of the disease. Trigeminal paræsthesia may also occur.

6. *Abducens nerve:* limitation of the mobility of the eye externally, due to involvement of the external rectus; deviation of the eyeball internally, and simultaneous diplopia when the eyes are directed outward.

7. *Facial nerve:* paralysis of the mimetic muscles of the face (upper and lower facial), of the mouth, the nose, the orbicularis oculi, and the forehead. The face-wrinkles become obliterated, closing of the lids is impossible (lagophthalmos), inability to pucker the mouth in whistling and to draw back the angle of the mouth in laughing.

Paralysis of the bellies of the stylo-hyoid occurs without symptoms.

If the seat of the lesion is the nerve in its course in the petrous portion of the temporal bone (Fallopian canal), there may result impairment of the chorda tympani from the second (or third [?]) branch of the trigeminus. Thus may arise disturbances of the sense of taste in the anterior portion of the tongue.

As irritation symptoms there may result spasm of these muscles, facial convulsion (tic convulsif); blepharospasm can also be produced in this way either reflexly or by central involvement.

8. *Auditory nerve.*

a. Cochlear nerve: deafness, paræsthesia.

b. Vestibular nerve: disturbance of equilibrium, attacks of vertigo associated with severe ringing and splashing sensations in the ears, whistling noises, and vomiting, thus constituting the symptom complex known as Ménière's disease.

9. *Glossopharyngeal nerve:* partial destruction of the sense of taste in the posterior section of the tongue. Anæsthesia of the mouth and of the œsophagus.

10. *Pneumogastric:*

a. Sensory branch (entirely vagus): anæsthesia of the throat, larynx, œsophagus, trachea, and bronchi.

b. Motor branch (in part from spinal accessory): paralysis of swallowing, paralysis of the œsophagus, disturbance of the function of the stomach, heart, and respiration. Paralysis of the recurrent laryngeal, paralysis of the vocal cords, and when the lesion is double-sided there will be aphonia.

11. *Spinal accessory:*

a. Its vagus recurrent branch (inferior laryngeal): palate and fauces paralysis.

b. External branch: paralysis of the sterno-cleido-mastoid and in part of the trapezius. Inability to draw the head to the side and to shrug the shoulders. As irritation symptom there will be spasmodic torticollis.

12. *Hypoglossal nerve:* paralysis of the tongue on one side, deviation of the tongue toward the paralyzed side on efforts to protrude it, articulatory speech disturbance, defective formation of letters.

III. Lesions of the More Important Spinal Nerves.

For details which cannot here be considered, the reader is referred to Plate 23 and the accompanying text. The distribution of sensory disturbance which is not mentioned can be seen in Figs. 11 and 12.

1. Occipitalis major nerve: occipital neuralgia, anæsthesia (see Fig. 10, *oma*).

2. Phrenic nerve: paralysis of the diaphragm and, as an irritation symptom, contraction of the diaphragm manifested by singultus.

3. Circumflex nerve: paralysis of the deltoid, inability to raise and abduct the arm, anæsthesia, see Fig. 11.

4. Posterior thoracic nerve: paralysis of the serratus, the shoulder stands out from the thorax, inability to raise the arm from the side to a horizontal.

5. Anterior thoracic nerve: paralysis of the pectoralis major, inability to abduct the arm.

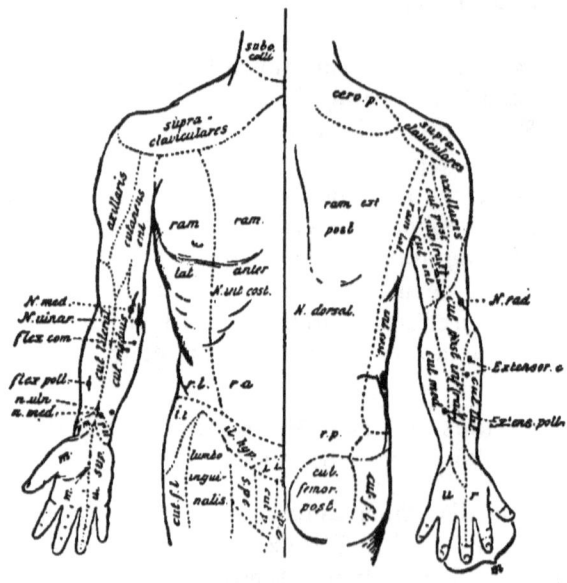

Fig. 11.

6. Musculo-cutaneous nerve: paralysis of the biceps and internal half of brachialis anticus (flexors of the forearm) and anæsthesia over lateral cutaneous nerve, for the distribution of which see Fig. 11.

7. Radial nerve [musculo-spiral nerve]: paralysis of the triceps (extensor of the forearm) and brachioradialis, absence of the prominence of these muscles on flexion of the forearm. Paralysis of the extensors

of the hand and proximal phalanges of the fingers and the extensors and abductors of the thumb. For the resulting anæsthesia see Fig. 11, the posterior superior and inferior cutaneous branches. The hands

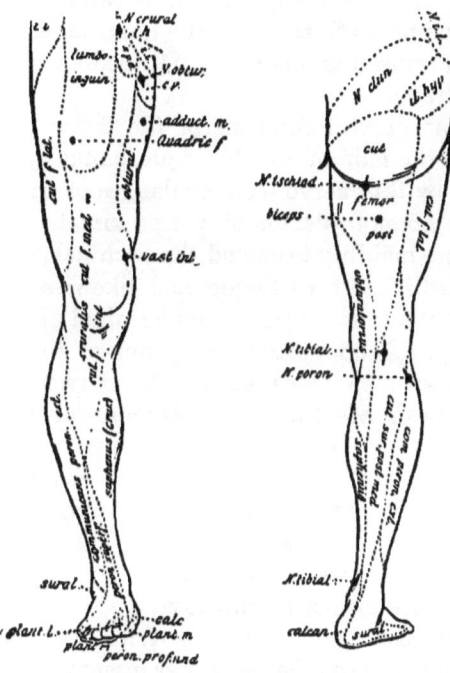

FIG. 12.

and fingers hang in a paralyzed attitude and the grip of the hand is very much weakened from paralysis of the antagonistic muscles.

8. Median nerve: paralysis of the pronators in flexed position of the forearm, and of the flexor muscles of the hand and terminal phalanges of the fin-

gers and thumb on the radial side, that is, all except the flexor carpi ulnaris and the ulnar half of the flexor profundis digitorum. Paralysis of the antagonistic muscles of the thumb and of the extensors of the end phalanges of the second, third, more rarely the fourth fingers due to involvement of the lumbricales.

For the anæsthesia see Fig. 11, median and volar median branches.

9. Ulnar nerve: paralysis of the flexor carpi ulnaris and the half of the flexor profundus digitorum supplied by this nerve (end phalanges of the fourth and fifth fingers), flexion of the proximal phalanges impossible, inability to extend the terminal phalanges of the fourth and fifth fingers and likewise inability to adduct the thumb and flex its terminal phalanx.

Through the unopposed contraction of the antagonistic muscles the so-called claw-hand results. For the anæsthesia see Fig. 11, superficial and posterior ulnar.

In paralysis of the ulnar, median, and musculospiral, there is great difficulty in writing, in grasping and holding an object, etc. In complete paralysis of only one nerve the hand is greatly handicapped in the performance of its function.

10. Intercostal nerve: sensory irritative symptom, intercostal neuralgia, herpes zoster intercostalis. Anæsthesia.

11. Crural nerve: paralysis of the psoas and quadriceps femoris, inability to raise the thigh and extend the leg. Inability to walk, to rise from the sitting posture, absence of patellar reflex. For anæsthesia see Fig. 12, cutaneus femoris, internal and median and saphenus nerve.

12. Obturator nerve: paralysis of the adductors of the thigh. (Anæsthesia, see Fig. 12.)

13. Sciatic nerve: complete paralysis of the feet and toes, as well as inability to flex the leg. Anæsthesia (see Fig. 12) of peroneal nerve, sural, cutaneous, and plantar nerves. Irritative symptoms, neuralgia of the sciatic (sciatica) and its branches.

14. Peroneal nerve: paralysis of the peroneal muscles, tibialis anticus, extensor longus digitorum and extensor hallucis; dorsal flexion of the foot impossible, foot drop, the outer border of the foot hangs lower than the inner.

On account of contracture of the oppposing muscles (calf muscles) there develops pes equinus and eventually pes varus when only the extensor digitorum is involved. Anæsthesia of the peroneal (see Fig. 12).

15. Tibial nerve: paralysis of the calf muscles, inability to perform plantar flexion and to flex the toes.

Development of pes calcaneus, through contracture in the opposing muscles (peroneal and extensors). Anæsthesia, see Fig. 12.

D. SYMPTOMS OF DISEASE OF THE SYMPATHETIC.

Paralysis of the cervical portion causes myosis, narrowing of the aperture of the lids, retrocession of the eyeball (oculo-pupillary fibres from the first dorsal segment), and anomalies of sweat secretion on the same side of the body. As irritation symptoms there may be dilatation of the pupil and disturbances of circulation.

From affection of the cardiac and splanchnic

branches there may arise disturbance of the heart, stomach, and glandular activity, but the actual symptoms are not well known. It would appear that such disturbance is very largely motor, possibly somewhat secretory.

II. Symptomatology of the System Diseases.

By the term system diseases we understand, as has been said above, such processes as involve symmetrically lying tracts or neurons which are functionally and anatomically connected, or (in the beginning) parts of neurons. Eventually the disease does not confine itself to a portion of a neuron (cell or fibre) but the entire neuron becomes the seat of disease. The localization of the primary seat of the disease is frequently of great importance in pointing the clinical course. Concerning the direction in which this process extends and as to its point of origin there is yet a great deal of obscurity.

1. System Diseases of the Cortico-Muscular Tract: the Motorial Pathway.

a. Of the central neuron: purely spastic spinal paralysis. Cases of this kind are of rare occurrence. Anatomically there is a slow, progressive degeneration of the pyramidal tract in its entire course, the starting-point being the pyramidal cells of the cortex (?). Clinically there is a slowly developing spastic paralysis of the extremities, with lively reflexes, without muscle atrophy and without disturbances of sensibility or of the bladder.

b. Of the peripheral neuron: there results a flaccid paralysis of the muscles with degenerative muscle atrophy. The reflexes are lost.

The primary seat may be either in the neuron cell or in the neuron fibre.

α. Primary diseases of the cell. These are divided according to the seat of the lesion into disease processes which have very different symptoms.

Chronic and acute ophthalmoplegia (seat of lesion, the nuclei for the eye muscles).

Bulbar paralysis, acute and chronic (seat of lesion, the bulbar nuclei of the facial, the hypoglossal, and trigeminal nerves).

Spinal muscular atrophy (seat of lesion, the gray anterior horns, especially of the cervical portion).

Poliomyelitis anterior, acute and chronic. The inflammatory process is in the gray anterior horns and the ganglionic cells situated there are very soon destroyed.

β. Primary disease of the fibre, neurotic muscular atrophy, peroneal type of progressive muscular atrophy (peroneal, ulnar, and median nerves). Involvement of the ganglion cells cannot be entirely excluded.

The motor form of multiple neuritis which may involve all the different motor nerves.

γ. Progressive muscular dystrophy. The disease confines itself to the terminal portions of the neurons in the muscles.

c. The entire tract: in this there is a degeneration of the pyramidal tract and the peripheral neuron, the ganglion cell, the outgoing fibre, and the peripheral nerve, thus constituting amyotrophic lateral sclerosis (eventually bulbar paralysis). The clinical symptoms are those of a spastic paralysis of the extremities with degenerative muscular atrophy.

In all probability there exist between these various forms others not yet clearly differentiated.

2. *The System Diseases of the Centripetal Sensory Tract.*

In this category isolated disease of the central tract is unknown. Nevertheless there occurs a striking degeneration of the peripheral neuron in tabes dorsalis, in which the neuron in its entire extent may be involved. Much more rare is disease of the central tract, although it may occur, such as degeneration of the lateral cerebellar tract and the anterior ground bundles. The symptoms consist of painful sensitiveness, disturbance of sensibility and of co-ordination, and loss of reflexes. To this division belongs the sensory form of multiple neuritis.

3. *The Combined System Diseases.*

They consist in simultaneous degeneration of the motor and sensory tracts. To this category belong:

a. Hereditary ataxia, Friedreich's tabes. In this disease the central motor neuron (pyramidal tract), in addition to the peripheral and central sensory neuron (posterior columns and direct cerebellar tracts), are involved. For symptomatology see special portion.

b. Combined tabes, genuine tabes with involvement of the lateral columns (disease of the pyramidal tract); and likewise other portions, *e.g.*, the peripheral motor neurons, may be diseased at the same time (motor oculi, abducens paralysis).

c. A number of other processes, anatomically somewhat better understood than clinically, such as disease of the posterior tracts, the direct cerebellar tracts, the pyramidal tracts.

Whether or not other (brain) tracts may be primarily systemically diseased is not yet known.

4. GENERAL REMARKS CONCERNING THE MODE AND METHOD OF EXAMINATION, TOGETHER WITH A DIAGNOSTIC SURVEY.

A. The examination follows the taking of the patient's history. The latter is of the greatest importance in leading to a diagnosis.

Among others the most important points to be investigated are:

a. The heredity (family nervous diseases, parental consanguinity, psychoses, alcoholism, tuberculosis, syphilis).

b. The previous life of the patient.

Method of life and habits, method of education and bringing up, temperament, alcohol, tobacco, morphia, excesses of any form.

Occupation (strain, handwork, lead, arsenic).

Previous disease, psychical and bodily trauma, the acute infectious diseases, tuberculosis, syphilis, etc.

c. The present disease, its causation, origin, and course.

B. The examination of the nervous system, properly speaking, consists under all circumstances of a careful investigation of the general condition as well as an examination of the more important internal organs; the state of nutrition, the bodily strength, the condition of the lungs, the heart; the abdominal functions and an examination of the pulse and urine. The patient should be carefully observed for symptoms of lues, tuberculosis, purulent foci, malignant tumors, disease of bone, diabetes, and for affections of the eye and ear.

I. Examination of the Motor Sphere

involves (1) the external appearance of the muscles, atrophy, hypertrophy, muscular depressions, information of which can be obtained by inspection contrasted with the normal contour of the body. In this way is to be detected atrophy of the muscles of the hands, peroneal atrophy, facial paralysis, atrophy of the shoulder girdle, abnormal position of the extremities, etc.

Measurement of the atrophic extremity and comparison with the normal side comes under this heading.

(2) The motorial irritative symptoms. Under this are to be reckoned muscular twitching, tremor of the extremities (in alcoholism, morphine habit, neurasthenia, Basedow's disease, etc.). Two forms of tremor are differentiated: a rapid, about twelve per second; and a slow, about six oscillations per second.

A special form of tremor is that of paralysis agitans and the intention tremor of multiple sclerosis, which, however, should be looked upon more as a disturbance of co-ordination than a tremor. Nystagmus of the eye muscles belongs in this category.

Contractions (spasms) of individual muscles or muscle groups, occur as tonic (continuous) or clonic (interrupted) form. Tetanus is a tonic spasm of the muscles of the body. Contractures are permanent tonic spasm conditions, while convulsions are external convulsive manifestations generally of a clonic but also of a tonic form. Epileptiform spasms are periodically occurring convulsions and so likewise are hysterical spasms. Choreic movements are involuntary uncontrollable movements, not of a brusque order like true spasms, which occur even when the patient is quiet; athetoid movements are more or less rhythmical and purposive movements, generally of the fingers or toes. This latter condition occurs after localized disease of the cortex, apoplexies, etc., and also spontaneously.

Fibrillary muscullar twitchings have been considered above.

(3) The examination of motor conductivity (motorial strength). This is important for all the more important muscles. To measure the strength we may use the dynamometer, but in most instances it is to be estimated by comparison with the other hand, or with the grip of one's own hand.

The eye muscles are tested by examining *seriatim* the function of the different muscles in the order of their nerve supply (Plate 23 and the accompanying text). Active and (for the extremities) passive mobility are to be considered after the size and strength of the parts have been examined. The condition of the joints should be carefully observed.

Very important likewise is the gait, whether it is paretic, slow, due to weakness of the muscles; spastic, stiff, the result of muscular stiffness; ataxic, unsteady, due to disturbed co-ordination; hemiplegic with one leg dragging, abducted, slightly contracted, the result of central paralysis, or peroneal with "drop toe," the result of paralysis of the peroneal muscles.

(4) The examination for the power of co-ordination.

This is done by testing for some more or less complicated action, such as placing the finger on the tip of the nose, etc., for the arms, and by requesting the patient to place the heel of one foot on the opposite knee to test the legs. Disturbance in the certainty of purposive movements is called ataxia, and it occurs in multiple sclerosis, tabes, hereditary ataxia, multiple neuritis, etc. Static co-ordination is tested for by asking the patient to stand with the feet together when the eyes are closed. If swaying results, the symptom is known as Romberg's phenomenon. Cerebellar ataxia, a tumbling gait like that seen in a drunken person, is easily recognized.

(5) Examination for the electrical conditions of the muscles (electro-diagnosis).

Electricity is a diagnostic aid of the first rank.

One uses for the electrical examination the galvanic, primary, constant current and the faradic, induced, secondary, interrupted current. The galvanic is the more important.

a. The Galvanic Examination.

The necessary apparatus consists of

(1) Elements with two fluids.
- Bunsen's element. { Zinc in H_2SO_4 / Carbon in HNO_3
- Grove's element. { Zinc in H_2SO_4 / Platinum in HNO_3

(2) Elements with only one fluid.
- Leclanché's element. { Zinc / Manganese } in HCl
- Bunsen's element. { Zinc / Manganese } in $KMnO_4 + H_2SO_4$

with the addition of neutral sulphate of mercury to offset the used-up zinc amalgam.

From the chemical process going on in the elements (salt formation) there occurs at both poles (the zinc pole, oxygen pole, negative; and the carbon or the positive, hydrogen pole) a difference of electrical tension, for the equalizing of which a wire connection suffices.

The current flows from the positive pole (anode) to the negative pole (cathode), that is, from the carbon to the zinc (mnemonically C before Z). The electromotive force depends, independent of the resistance of the elements—that is, the internal resistance—upon the resistance which the external current has to overcome. The dry human skin forms, if it be interposed in the circuit, at first a very high resistance, which, however, diminishes as the current continues to pass until it finally remains constant.

The intensity of the current is measured by the galvanometer. It is expressed on the absolute galvanometer in milliampères (MA), an arbitrarily chosen, now universally used unit of measure.

The intensity of the current can be graduated by means of the rheostat, that is, by the interposition of resistance coils.

At every excitation with the galvanic current the healthy muscle reacts by contracting with the closure of the current as well as with the opening. These contractions are characterized by their quick beginning and lightning-like rapidity. On application and increase of the current contractions occur in the following series (formula of contraction): Cathodal closure (KaCC) first, then with a stronger current they occur with anodal closure (AnCC); then follows a contraction with the anodal opening AnOC, and last with a stronger current a contraction with cathodal opening, KOC.

If the current be made stronger there develops a tetanic contraction, KCTe before AnCTe.

The response of degenerative muscles is quite different. In these the contractions are not quick and lightning-like, but slow, indifferent, vermiform, wave-like which may be followed by the eye. At the same time AnCC occurs before KaCC.

This perverted response is known as the reaction of degeneration.

As degenerative muscular atrophy follows lesion of the peripheral motor neuron alone, we have in the presence of reaction of degeneration a diagnostic factor of paramount importance in differentiating disease of the central from disease of the peripheral motor neurons. Reaction of degeneration is present if the cell or the fibre of the peripheral motor neuron is diseased, and is not present if they are not the seat of disease.

A complete and partial reaction of degeneration is recognized.

In complete reaction of degeneration it is impossible to cause contraction of the muscle by irritating its motor nerve (indirect irritation). Possibility to excite the muscle directly (direct irritation) is the next to be lost in reaction of degeneration. If the degenerative atrophy has gone on to a marked degree and the muscle has been transformed into a mass of connective tissue, direct irritability of the muscles will be lost; if muscular tissue eventually returns, the reaction of degeneration disappears and the contractions become normal and the muscle becomes irritable both through the nerve and directly.

Partial reaction of degeneration is said to exist when excitation of the muscle through the nerve is not lost but direct irritability is gone. This condition is the result of a very much less severe injury than complete reaction of degeneration.

We have, therefore, in the galvanic examination not only the means of determining the locality of the disturbance (peripheral neuron) but of the severity of the disease process as well—that is, the prognosis.

If in a few days after the development of a lesion of the peripheral neuron which is accompanied by muscular paralysis there does not develop reaction of degeneration, it may be assumed that we are dealing with a mild form of the disease which will disappear in a comparatively short time. If partial reaction of degeneration develops it is safe to say that the duration of the paralysis will be from weeks to months; while if complete reaction of degeneration develops several months will elapse (four to nine) before recovery begins to show itself, and it must be said that it may never set in; that is, the paralysis may remain.

In all nuclear lesions of the peripheral neurons (spinal muscular atrophy, poliomyelitis anterior, amyotrophic lateral sclerosis, myelitis) as well as all severe defects or lesions of the anterior motor roots and the peripheral nerves (neuritis, compression, trauma, etc.), reaction of degeneration is accompanied by fibrillary twitchings and atrophy.

With these qualitative changes there may develop

quantitative diminution or increase of muscular irritability, but quantitative changes have not the pathognomonic significance that qualitative have.

Before speaking of these changes we wish first to refer briefly to

b. The Faradic Examination.

The customary faradic (induction) apparatus consists, in addition to the necessary primary current which is produced by a galvanic element (generally a modified Bunsen or Leclanché cell), of a Du Bois-Reymond sliding inductorium, *i.e.*, a sliding roll of wire, an induction coil. The primary current passes through the primary coil and excites by induction in the secondary coil which is wound about the primary, a current which runs in the opposite direction.

Since it is a fact that by an interposed interrupter, for instance a Wagner's hammer, the primary current is continually interrupted and its direction reversed (electro-magnetic action), then secondary currents are induced in the induction coil in the reversed direction (also alternating), the action of which is strengthened or weakened according as the secondary coil approaches or recedes from the primary. The intensity of the current may be read off in centimetres from the scale attached to the sliding coil. The greater the distance in centimetres the weaker the current.

With the changes in qualitative reaction to the galvanic current which constitutes reaction of degeneration, faradic irritability is always found to be lost. This is shown as well when the current is applied to the nerves as when applied directly to the muscles. When irritability returns, the response to the galvanic current precedes that for the faradic. In reference to the relative quantitative irritability it is necessary to bear in mind that different nerves and muscles require very different currents to produce a contraction, according to the position and make-up of the different structures. Therefore in order to be able to demonstrate quantitative pathological changes one must be familiar with the normal averages. These are determined for the different nerves and muscles at certain points, the motor points. Only electrodes of certain size should be used, the normal electrode being 3 sq. cm. The most important points for nerve trunks and muscles are to be seen in Figs. 10 to 12, p. 102 *et seq.* The most important averages for the normal galvanic irritability in MA. and the faradic in cm. on the scale attached to the sliding coil are, for the nerves (muscle averages are of less importance), the following, taken from Stintzing.

	Galvanic. MA.	Faradic. CM.
Facial nerve	1.75	121
Frontal branch	1.45	128
Mental branch	0.95	132
Accessory nerve	0.27	137
Median nerve (in arm)	0.9	122
Ulnar nerve	0.55	130
Musculo-spiral nerve	1.8	105
Crural nerve	1.05	111
Peroneal nerve	1.1	115

A pathological increase of galvanic irritability is found especially with reaction of degeneration; and likewise apart from reaction of degeneration in some diseases, such as tetany, myelitis, etc. (also excessive faradic irritability). There is decrease of electrical irritability of all kinds in simple atrophy (central paralysis).

Pathologically increased resistance to the conduction of the electrical current is found in Basedow's disease and in conditions where there is diminished moisture of the skin.

II. Examination of the Sensory Sphere.

1. Investigation of subjective complaints made by the patient such as sensitiveness to pain, its localization and kind, and likewise of abnormal sensations or paræsthesias, such as burning, feeling of fur, formication, conditions which are found with neuralgia, neuritis, tabes, and myelitis.

2. Objective examination to test the preservation of the different qualities of cutaneous sensibility.

a. *Tactile sense* is tested at the same time as the sense of pressure.

In making these tests one requests the patient to differentiate between the hand and the point of a pin, to differentiate between a bit of cotton wool and the end of the finger, and the moment contact is felt to say "now." As a control one should make false touches and observe if the patient answers. Particular care should be taken not to exhaust the patient by such examinations. Frequent repetition of this examination at different times is absolutely essential in order that safe conclusions may be drawn. When it is possible the sensitiveness of the disordered side should be compared with the healthy side and the entire cutaneous region should be examined. Likewise cutaneous sensitiveness to the faradic current should be investigated. There may be anæsthesia, entire loss of the sense of touch (in total interruption of conductivity especially by focal diseases, see this Section, 3, and also in hysteria); hypæsthesia, a partial loss of tactile sense caused by partial interruption of conductivity such as in neuritis, tabes, etc.; and hyperæsthesia, a pathological increase of sensitiveness (see under Sensitiveness to pain). The extent of an anæsthetic area can be made out with exactness; it corresponds to the distribution of one or more peripheral sensory nerves. It signifies lesion of a peripheral nerve and its ramifications (see Figs. 10, 11, and 12).

b. Localization of Sensations (tactile sensibility). This is tested simultaneously with the sense of touch. The patient should tell the localization of a touch when the eyes are closed.

The ability to recognize two points as such is tested by a pair of calipers. The following table, taken from Weber, indicates the shortest distances and the parts of the body on which two points can be recognized.

	Millimetres.		Millimetres.
Tip of the tongue	1	Upper arm	65
Mucous membranes of lips	4	Forearm	39
Tip of the nose	6.5	Hand { volar side	11
Skin of cheek	11	{ dorsal side	28
Forehead	30	Finger tips	2
Skin of chest	44	Thigh	65
Middle of back	65	Leg	39
Buttocks	39	Tip of toes	11

The less the distance in which two points can be recognized the greater is the sensibility in that part.

In polyæsthesia simple contact will produce multiform sensations (tabes).

c. *Pain Sensibility.*

This is tested by needle pricks of varying severity on the individual parts of the body. [Simple pinching of the skin and deeper tissues between the ends of the fingers is often more serviceable than needle pricks.] A measurable test for pain we have in cutaneous faradization, the current being increased until it produces painful sensations, then the amount is read off on the scale. This is only necessary for the more refined clinical examinations.

Sense of pain may be diminished (hypalgesia) or it may be entirely lost (analgesia).

Analgesia is associated with anæsthesia to touch in hysteria, tabes, neuritis, focal diseases, and without anæsthesia in syringomyelia.

Delayed sensibility to pain, that is, recognized only when there is a summation of stimulation [long continuation] is often present in tabes, myelitis, neuritis. In testing for this it will be noticed that the patient will first indicate the time of contact and some time afterward the time of the painful sensation, *e.g.*, "now . . . . . . ugh."

Abnormal continuance of pain sensations (lingering) is also sometimes found in tabes and in neuritis.

Hyperalgesia, abnormal sensitiveness to contact and to heat and cold, is present in focal diseases as a symptom of irritation, in hysteria, neurasthenia, in unilateral lesions, and in neuritis.

d. *Temperature Sensibility.*

Testing for sensibility to heat and cold should be done with test tubes, one filled with warm (not hot) the other with ice water.

Sensibility to heat and cold may be diminished or increased. There may be thermo-anæsthesia for both or only for one in focal diseases, tabes, syringomyelia, myelitis, neuritis. The two sensations may be transposed, cold for warm and vice versa, *i.e.*, perverse temperature sensibility sometimes seen in tabes, myelitis, and neuritis.

3. SENSIBILITY OF THE DEEPER PARTS.

a. Sensibility to weight is tested by placing objects of different weight on different parts of the body, such as the hands, and comparing the impression that it makes on the patient with the normal side or with a normal individual.

b. Sensibility to passive motion is tested by performing some small excursive movements on the patient. With closed eyes he is requested to state the flexion, extension, raising, and lateral bending of the various joints.

c. Sense of position is tested by requesting the patient to put the hand, the arm, the leg, etc., in a position corresponding to its fellow when the eyes are closed.

Disturbance of these and other sensations more difficult to test, such as the articular sense, muscular sense, etc., are classified together under the caption of disturbance of the "muscle sense." They are to be sought for in hysteria, tabes, hereditary ataxia, neuritis, and in focal diseases.

4. THE HIGHER SENSIBILITY.

a. *The Sense of Sight.*

Each eye and its homologous half of the retina is to be carefully examined.

The exact determination of visual acuity and of the visual field by the perimeter is often necessary (see text-books on diseases of the eye).

Casual orientation of the state of the visual field can be obtained by placing the fingers or a small bit of white paper beyond the visual field and then gradually bringing them into it, vision being directed to a point straight ahead. If, for instance, on the left side (of the patient) the finger is not seen until it approaches the middle, while on the right side it is recognized at a considerable distance outward, one may be sure that the patient has left-sided hemianopsia, bilateral and homonymous—an affection of either the optic tract of the right side, or of the primary optic centres, the optic radiation of the occipital lobes of the right side.

In a similar way one is able to recognize concentric limitation of the field of vision or central defect in the visual field, scotomata. Loss of vision in the temporal fields, hemianopsia, points to affection of the chiasm.

If the optic nerve is affected, there will exist a more or less severe amblyopia, eventually amaurosis of the affected eye. Diplopia points to affection of the muscular apparatus of the eye.

It is indispensable in many nervous diseases to make a careful ophthalmoscopic examination. In this way most important information, such as choked discs (a prominent forward projection of the papillæ with ill-defined border) may be obtained. Choked disc arises sequentially to impediment to the return lymph circulation in the sheath of the optic nerve, and by increase of the intracranial pressure, as from tumor. Similar optic neuritis with slighter exudation accompanies multiple neuritis, brain syphilis, meningitis, etc. This may eventuate in atrophy of the optic nerve, an atrophy which is manifest by a white or pale papilla and a sharply defined border.

Optic atrophy may develop primarily with tabes, multiple sclerosis, and dementia paralytica.

b. *Sense of Hearing.*

Each ear should be tested separately with the watch, with the voice at a whisper, and with sounds at greater distance.

In order to determine if the difficulty of hearing is dependent upon labyrinthine affection or affection of the bone conductivity to the middle ear the tuning-fork is made use of. If, when there is difficulty of hearing, a vibrating tuning-fork is placed against the skull and the sound is not heard, *i.e.*, through bone conductivity,

(*a*) more acutely than if held before the ear, this constitutes a positive Rinné, a condition which bespeaks disease of the labyrinth (meningitis, disease of the auditory nerve);

(*b*) Or if it is not heard here, negative Rinné, then it indicates disease of the middle ear.

Bone conductivity is entirely lost in nervous disturbance of a severe degree.

Under auditory paræsthesia may be reckoned sowsing and ringing in the ears, which accompany otitis, Ménière's disease, anæmia, etc. Auditory hyperæsthesia occurs in hysteria, hemicrania, etc.

c. The Sense of Smell.

Each side of the nose should be tested by placing before it some odoriferous substance, but which does not, like ammonia, irritate the terminal ramifications of the trigeminus. For this purpose oil of peppermint, oil of cloves, or asafœtida may be used. Loss of the sense of smell, anosmia, occurs, aside from disease of the nasal mucous membrane, in peripheral lesions, such as tumors, injury to the anterior fossa of the skull, atrophy of the olfactory nerve, with central disturbances, and also with hysteria.

d. Sense of Taste.

Each side of the tongue should be tested in its anterior two-thirds (trigeminal), and its posterior third (glosso-pharyngeal). These should be tested separately by applying a bit of salt solution, sugar, vinegar, and quinine; thus testing for the salt taste, the sweet, the sour, and the bitter. Loss of the sense of taste (ageusia) accompanies peripheral nerve lesions, lesion of the facial nerve in the temporal bone, lesion of the trigeminal, and hysteria, but it is seldom found with disease of central origin.

III. Investigation of the Reflexes.

The reflexes may be increased, diminished, or lost. Increase of the reflexes may be brought about by a withholding of the normal inhibitory impulses through the cerebral pathways (as in fatigue, neurasthenia, disease of the inhibitory pathways) or by abnormal irritability within the reflex arc (neuritis, meningitis, tetanus).

Diminution and loss of a reflex may result from interruption of some portion of the reflex arc (the sensory or the motor portion, or the communicating portions of the reflex collaterals). The reflexes are lost in deep coma and in intoxication. When the tendon reflexes are greatly increased there occur on continual irritation reflex contractions or twitchings following rapidly one on the other, thus constituting a patellar clonus, ankle clonus, or the so-called foot phenomenon.

The most important reflexes to examine are:

1. The skin reflex and the reflex of mucous membranes.

The skin reflex of the sole of the foot and the points of the fingers are tested by means of needle pricks, by cold and warm, and by tickling. Their pathways are very little known; their sensory branch must be in extensive connection through collaterals and reflex neurons with the entire motor sphere.

One-sided foci which produce interruption of the pyramidal tract cause increase of the tendon reflexes, while the skin reflexes are frequently lost.

The cremaster reflex, contraction of the cremaster muscle, brought about by irritation of the skin on the inner surface of the thigh, and

The abdominal reflex, contraction of abdominal muscles at different heights, upper and lower abdominal reflexes, caused by stroking the skin of the abdomen, are of comparatively slight diagnostic significance. Both of these reflexes appear to be lost with focal areas of disease in the brain, on the side opposite to the lesion, and on both sides in multiple sclerosis.

The conjunctival reflex, which causes closure of the lids on irritation of the conjunctiva, is lost in paralysis of the fifth and seventh nerves.

The winking reflex, which causes closure of the lid when anything is suddenly brought before the eyes, is lost in lesions of the optic nerve and paralysis of the seventh. The pharyngeal reflex, brought about by irritation of the mucous membrane by a small brush or a bit of cotton, and which causes a choking movement, is lost in paralysis of the vagus and accessorius. The palate reflex causes a contraction and elevation of the palate.

2. The periosteal and tendon reflexes.

(*a*) The patellar tendon reflex is the most important. [Tendon reflex is a glaring misnomer. The reflex effect is not in the tendon but in the muscle. The expressive terms, knee jerk, ankle jerk, etc., are to be preferred.] Tapping of the patellar tendon is followed by a contraction in the quadriceps extensor, which causes the foot and leg to rise quickly into the air. The vigor and intensity of this contraction may be increased by what is called reinforcing the reflex (Jendrassik's method). In order to do this the attention of the patient is occupied by asking him to pull the clenched hands at the moment when the tendon is struck, or to squeeze the closed eyelids tightly together at that time, etc. The patellar reflexes are best tested by having the patient sit on a moderately low seat, the feet squarely on the floor and the legs slightly forward, forming a very obtuse angle with the thighs.

This reflex differs physiologically in different individuals. Increase of the reflex is the rule in neurasthenia, in lesions of the central motor neurons (reflex inhibitory fibres in the lateral pyramidal tracts), in any condition which causes irritation of the reflex arc, as meningitis, tetanus, neuritis, and in exhausting diseases.

Loss of the reflexes occurs whenever the reflex arc is interrputed as in tabes, neuritis (crural), anterior poliomyelitis, lumbar myelitis, in coma, epileptic attacks, after recent spinal-cord injuries (irritative symptom of the inhibitory fibres?).

(*b*) Similar conditions prevent the occurrence of the Achilles reflex, a phenomenon which consists in a contraction of the calf muscles on tapping of the Achilles tendon. Exaggeration of this constitutes foot [ankle] clonus, the foot phenomenon which is produced by forcible and sharp dorsal flexion of the foot.

Of the periosteal and tendon reflexes of the arm, the following are the most important:

The radial and ulnar periosteal reflex, a contraction following tapping of the styloid process of the radius or the ulna; the triceps tendon reflex, produced by a smart blow on its tendon just above the olecranon process.

Aside from these the masseter reflex deserves investigation. This is produced by tapping the chin when the mouth is partly opened, the result is a quick, partial, or complete closure of the mouth.

The clinical significance of the reflexes considered under this caption is not very great; the pathological changes which they attend being practically those mentioned above.

3. The pupillary reflexes.

The following subdivision is made.

1. Light reaction; narrowing of the pupil of the same eye, by contraction of the sphincter on exposure to light (irritation of the optic nerve).

2. Consensual light reaction, contraction in one eye when the other is exposed to light.

3. Reaction to accommodation, narrowing of the pupils on accommodating for near objects, essentially an associative movement.

4. Convergence reaction, of very slight clinical significance.

The light reaction is tested separately for each eye (to eliminate consensual reaction) and in cases of hemianopsia likewise the impressionable halves of the retina. This is best done by means of reflected light, ordinarily enough shade is given by the hands; then to get the reflex of the one eye, the hand covering is taken away quickly.

Before the test is made one should observe carefully any abnormal dilatation or contraction of the pupils.

Enlargement of the pupils, mydriasis, occurs with atropine and cocaine poisoning, in blindness, coma, during an epileptic attack, and with paralysis of the motor oculi, etc.

Narrowing of the pupil, myosis, occurs in morphine poisoning, tabes, dementia paralytica, meningitis, affection of the first dorsal segment, iritis, etc.

Inequality of the pupils is to be observed with paralysis, tabes, meningitis, etc.

The light reflex may be abnormally slow or lost when there is an interruption of continuity of the reflex arc, as in blindness from disease of the optic nerve, paralysis of the oculomotor, paralysis or lesion of the reflex collaterals of the oculomotor (tabes, dementia paralytica), in the disease of corpora quadrigemina, and aside from this in coma, narcosis, and during an epileptic attack, but not in hysteria.

In lesion of the optic tract there is hemianopsic pupillary stiffness (no contraction when the light falls on the blind half of the retina).

IV. Investigation of the Function of the Bladder and Rectum.

a. Lesions of the central tracts, dorsal myelitis, focal disease of the cord.

Lesion of the motor tracts causes retention of urine and of feces, voluntary evacuation is impossible, and after the bladder gets full the urine comes away in drops.

Lesion of the sensory tracts destroys the sensitiveness of the bladder, there is loss of urine pressure on the bladder, and consequent retention.

b. Lesion of the peripheral tracts and of the bladder centres (the reflex collaterals in the sacral cord) is attended by incontinence of urine and frequently of fæces, paralysis of the sphincter. Lesion of the lumbar sacral cord and of the cauda equina causes continual dribbling of urine. In this condition there is also inability to feel any sphincter contraction when the finger is forced through the sphincter, a condition which is preserved when the disturbance is central.

In addition to these paralytic appearances there may be reflex symptoms and symptoms of central irritation, such as slow urination, tenesmus, strangury, etc.

After primary disease of the spinal cord there develops very easily and very rapidly severe cystitis, which is quickly followed by secondary infection, pyelonephritis, and pyæmia.

V. Examination for Trophic and Vasomotor Disturbances.

Trophic disturbances are associated with lesion of the anterior horns, neuritis, disease of the blood-vessels, etc. The most important are:

Redness, swelling, cyanosis, abnormal paleness, urticarial eruption on the skin, erythromelalgia (painful redness and swelling of the hands and feet), angio-neurotic œdema (periodical or intermittent multiple areas of skin œdema).

Atrophy of the skin (glossy skin, shining skin), scleroderma, unilateral atrophy of the face (hemiatrophia facialis), anomalies of perspiration such as hyperidrosis, secretion of sweat on one-half of the body as in hysteria and neurasthenia.

Spontaneous formation of gangrene of the extremities, as Raynaud's disease, Morvan's disease, syringomyelia, bed-sores in paraplegia (only of indirect nervous origin); perforating ulcer in tabes, change in the joint, arthropathies, such as the swelling, thickening, proliferation of tabes, etc.

VI. Examination of the Psychical Functions.

1. Speech and Writing.

a. Articulatory disturbance. Dysarthria.

These are disturbances of speech due to defect in the peripheral tracts. It may be that only the utterance is defective for individual letters or the defect may be so great that the letters cannot be formed at all. The words come from the mouth in a choking, nasal, running-together fashion which may be entirely incomprehensible and which are produced by great muscular effort. This variety of disturbance of articulation is seen in bulbar paralysis, in lesion of the nuclei of hypoglossal and facial, in defective formation of the articulatory muscle apparatus, such as cleft palate, etc.

Special forms have been differentiated as typical of certain lesions, for instance, bulbar speech, nasal speech (rhinolalia), slow speech (bradylalia).

Scanning speech, syllabic splitting up of word formation, is considered typical of multiple sclerosis.

Various defects of speech formation, especially defects in formation of letters, are classified under stammering speech (dysarthria literalis).

Stuttering arises from abnormally strong spastic contractions in the speech muscles (of central origin?). Psychical excitation increases stuttering, dissipates stammering.

In dysarthric disturbances, examination is made of the lips, the tongue, the palate, and the nasal passages.

b. Aphasic speech disturbances.

These are conditioned through central disease and may be of a cortical nature (cortical foci), of transcortical origin (disturbance of the association tracts), and of a subcortical form (interruption of the conducting path from the centre to the periphery).

(1) If the patient cannot speak voluntarily but can understand everything that is said to him, then the speech disturbance is known as (motor-ataxic) aphasia. Lesion as in II., Fig. 9; for explanation see page 71.

If the patient cannot speak a single word voluntarily, he is said to have total motor aphasia.

If he can say only individual words, such as yes and no, etc., he is said to have monophasia.

(2) If the comprehension of words fails him without disturbance of speech, the patient is said to have sensory aphasia, word deafness. Lesion in I. In this condition there is a disturbance in the proper association necessary to speech—*i.e.*, an abolition of the necessary association between I. and II. Speech is false, non-pertinent, wrong words are used, and there may be also defective sentence formation, a condition known as paraphasia. A form of this is literal paraphasia in which there is stumbling in the syllables, the placing of letters and syllables in wrong connections (in dementia paralytica). If there is loss of word memory so that the patient cannot quickly remember names, designations, etc., as formerly, although no comprehension of the word or its meaning is lost, this condition is known as amnesic aphasia.

(3) Inability to write voluntarily is often associated with motor aphasia and very frequently with

sensory aphasia. This condition is known as agraphia. It occurs as the result of disturbance with V. or of the associations between I., II., and V.; see paragraphia.

(4) With and without sensory aphasia there comes inability to read, without any particular disturbance of vision. This is known as alexia, and is the result of lesion of the inferior temporal lobes, the gyrus supramarginalis.

All of these conditions may occur together in widespread disease, as may complete motor and sensory aphasia.

(5) If the patient does not recognize familiar objects and their significance, there is said to be psychical blindness (focus in III.). He does not find for familiar objects the right word, and when motor aphasia is not present, this condition is called optic aphasia (disturbance of the association between III. and I., II.).

Disturbances in the associations between I., III., V. cause (similar to paraphasia) paragraphia; just as association disturbance between I., II., and the supramarginal gyrus causes paralexia.

Both of these conditions in combination with defective syllable formation occur frequently in dementia paralytica.

Disturbance of writing mostly of peripheric localization is known as dysgraphia (analogous to dysarthria); such conditions are tremulous writing in alcoholic and senile tremor, etc., ataxic writing in hereditary ataxia and multiple sclerosis.

Functional disturbance of the power to read, such as rapidly induced fatigue, is known as dyslexia.

Examination of these central speech and other disturbances can be done in the following way:

One begins the examination with a question, "Have you been sick long?" If there is no answer there may exist (excluding deafness and lack of psychical comprehension):

a. Deaf-mutism, which is to be recognized by the gesticulations which at once begin.

b. Total motor and sensory aphasia, which is corroborated by having to get the history from a third person.

If the patient nods his head and likewise answers other questions correctly by means of gesticulations or mimic movements without, however, being able to bring forth a word, it may be taken for granted that the patient has motor aphasia. If he does not respond at all to the question or if he does not do some simple thing that is asked of him such as to close the eyes, it may be concluded that he has sensory aphasia; but if all questions or requests are quickly and properly responded to by gestures or mimicry, this can be entirely excluded.

If the patient answers all questions with one and the same word, for instance, "no," "no," then he has monophasia.

If he answers incorrectly, especially if he misplaces words, syllables, or letters, then he has paraphasia. For exact corroboration of this the patient should be asked to repeat certain polysyllabic words such as electricity, artillery brigade, or other long words.

After this certain familiar objects are exhibited to the patient such as a match, a lead pencil, etc., and he is questioned as to their use. If the patient is unable to respond correctly either by words or gestures (and if sensory aphasia and disturbance of vision can be excluded), then there exists psychical blindness. If the patient recognizes the objects but cannot find the right name for them, there exists optic aphasia. Now the patient is requested to repeat (the above disturbances being excluded) after dictation words, sentences, and to count. If he quickly forgets what has been spoken or if he repeats them in a defective fashion, then he has amnesic aphasia. Then follows the examination of the power of writing.

Words and figures are dictated and the patient is requested to write them. If he does not write at all, and sensory aphasia can be excluded, agraphia exists; if the same word such as "no," "no" is written and rewritten, then he has monographia; or if the writing shows that syllables and letters have been left out, the defect is called paragraphia.

Amnesic disturbance of writing exists when the patient forgets quickly what has been dictated. In a similar way one examines for the capacity to write spontaneously and to copy.

To test the power of reading, the patient is asked to read printed or written figures and numbers. If this is apparently not possible and motor aphasia and disturbance of vision can be excluded, there exists alexia, monolexia, paralexia, depending upon the form of disturbance. It is to be noticed whether or not the patient comprehends what he may be able to read.

Amnesic alexia is caused through disturbance of intellect, rapid forgetting of recently seen letters and syllables. This is determined exactly by requesting the patient to place words together and to count single figures and ciphers.

2. *Mental Examination.*

Much has already been said of this in speaking of the disturbances of speech in connection with intellectual disturbance and in discussing amnesic aphasia, alexia, and agraphia.

Aside from these the patient should be examined as to early memories, first impressions and memories, and also the most recent ones, for instance, the details of his illness.

Examination for recent impressions is done by bringing out simple and complicated processes of association.

Disturbance of memory is known as amnesia; it frequently attends injuries of the skull, compression of the brain, foci of disease in the brain, and dementia. It is tested through the various associations, auditory, optic, tactile, etc. Its course cannot be considered further in this connection.

3. *Other Psychical Disturbances.*

It is of great importance to determine if consciousness is disturbed (coma, somnolence, stupidity); if the intelligence is normal or disturbed (dementia, idiocy); if pathological motor or sensory irritative conditions are present, such as delirium, state of excitement with limitation of consciousness, hallucinations, delusions, and morbid delusions of central origin; and further than this, if there are any systematized ideas, fixed or fluctuating, or obsessions.

Conditions of melancholia (pathological depression), mania (pathological exaltation), and hypochondria are to be looked after.

By the presence or absence of such symptoms one reaches a conclusion as to the psychical condition and as to the diagnosis.

C. THE MAKING OF THE DIAGNOSIS.

The diagnosis follows principally after the complete examination which has just been outlined and the taking of the patient's history.

The most important questions to be considered are the following:

1. Does the patient suffer from a nervous disease or is nervous disease only simulated by anæmia, tuberculosis, tænia, etc.?

2. Is the cause of the nervous manifestations that have been found some disease of another part of the body and can such disease be made responsible for them? Such, for instance, as arteriosclerosis, tumor, disease of the heart, lungs, blood, kidney, stomach, etc., or a disease such as diabetes.

3. Is the disease of a functional or an organic nature? This is a question that cannot be settled off hand and frequently it requires prolonged observation of the patient. Choked discs, degenerative muscular atrophy, reaction of degeneration, loss of patellar and pupillary reflexes are, for instance, attributes of organic disease. One must always take into careful consideration the genesis of the disease.

4. What is the seat of the postulated organic disease (see General Symptomatology, Section IV., 3) and is the disease a focal or a systemic one?

5. Of what form is the disease? One must consider its mode of onset, its apparent connection with other diseases, such as infectious disease, tumor, etc., the location of the lesion, and the attending symptoms, the presence of fever, cachexia, etc.

Very frequently it is impossible to answer these questions satisfactorily after one examination and often it is necessary to watch the disease process and its course for some time. In many cases the complete diagnosis remains in suspense while the probable or provisional diagnosis may be made. One cannot be too careful, without the weightiest data, in making such diagnosis as dementia paralytica, tabes, tumor of the brain, from one examination alone.

5. GENERAL REMARKS ON THE TREATMENT OF NERVOUS DISEASE.

The treatment of nervous disease requires more than does any other form of disease that the physician be grounded in psychology and capable of treating his patient from that standpoint. Unfortunately this side of the development of our physicians leaves much to be desired. A knowledge of the anatomical and clinical facts is by a long way not enough.

The treatment involves

1. THE PROPHYLAXIS.

In no other class of patients can so much ill health be prevented and so much misery through ulterior consequences be mitigated as in the cases of which we speak.

The individual points to be attended to are: The

advice of the physician against marriage of those individuals afflicted with severe mental and nervous disease and possibly of those handicapped through inheritance (hereditary systemic diseases, etc.). Continual observation by the physician of the nutrition and hygiene of children during the years of their development and the anticipation and prevention of sinister methods of living which include the abuse of alcohol, tea, tobacco, excesses of every form, the use of morphine (even by the physician), the prevention of mental strain, irregularity in work and diet, hygiene of places of employment and likewise observation of personal hygiene, under which must be included personal cleanliness, bathing, clothing, exercise, etc.

2. Causal Therapy.

Unfortunately this form of therapy has until now been possible in but a few diseases.

a. Intoxication conditions (from lead, arsenic, alcohol, morphine, cocaine) may be cured or at least very much bettered, for the first-named poisons, by interdicting the factory work through which they were obtained or by some regulation of it; for the last-mentioned by abstinence, or treatment in an institution. Sudden withdrawal of alcohol, morphine, and cocaine is often followed by disagreeable but not dangerous consequences. In alcoholism abstinence manifestations do not appear when stimulants are stopped, statements to the contrary notwithstanding, and although such symptoms do appear in abstinence from morphine and cocaine and although they make the patient very wretched, they rarely signify real danger. None of the so-called cures "without compulsion or pain" are of permanent benefit. The administration of large quantities of alcohol to counteract the symptoms of abstinence from morphine is a dangerous procedure and should be avoided.

It should not be understood that, in severe cases of morphine habit, within the first three or four days complete abstinence can be reached. On the contrary, during the first few days of treatment, or to ward off threatened collapse, a small quantity of morphine should be given.

b. All syphilitics, as soon as the diagnosis is made (or when the suspicion of syphilis is strong), should be subjected to an energetic inunction cure, 3 to 5 grams (3 ss.- 3 iss.) of gray ointment a day. Iodide of potassium may be administered simultaneously or later, 3 grams (45 grains) daily. Never should the attempt be made to treat syphilis with iodide of potassium alone.

Malarial neuralgia should be treated by the administration of quinine 8 to 20 grains daily.

c. The third class of diseases, the recovery of which we may facilitate by treating the condition directly, are those in which surgical interference is possible. The number of diseases that are amenable to this form of treatment has in recent times been much increased.

Aside from conditions dependent upon direct trauma to the nervous system, there may be placed in this category:

(1) Diseases of the brain which may be exactly localized, tumors, bone fragments, localized purulent areas when they are in or very near to the cortex.

Where purulent foci are suspected, trepannation and exploration are justifiable procedures.

In increased intracranial pressure, as in hydrocephalus, trepannation and tapping of the ventricles have been followed by beneficial results; as has also lumbar puncture for the same condition. Carious bones and carious foci in joints which may entail injurious consequences to the nervous system come under the caption of diseases subject to surgical treatment. Whether or not Jacksonian epilepsy is at all benefited permanently by surgical treatment is yet a matter of question, and one that we shall be much more capable of answering when the pathology of epilepsy is understood.

The results of operations for tumors has heretofore not been very encouraging, while operation for brain abscess has been followed by excellent results.

(2) Spinal-cord diseases.

Aside from those arising from direct trauma, the diseases that we subject to surgical treatment are tumor of the cord and caries of the spinal vertebræ. In these, results up to this time have not been very encouraging. The uncertainty of diagnosis as well as the difficulty in its exact localization, the difficulty of orientating one's self in the field of operation, and other factors serve to militate against the success of surgical methods in these diseases. Nevertheless it may be said that the earlier the operation is undertaken in these cases, before the disease has produced irreparable changes in the cord, such as secondary degenerations, the greater may be one's anticipation of benefit.

When such degeneration has positively taken place

the beneficial results of any operation, no matter how successful it may be, must remain very slight.

(3) Peripheral nerve diseases.

Here the best results are got, because the power of regeneration in these parts is markedly better than in the brain or spinal cord. Even after a long time has elapsed since the continuity of the nerves has been severed, entire reparation of the nerve is possible.

The recovery of a nerve whose continuity has been severed may be hastened by plastic operation.

The most important factor leading to success in any of the conditions that allow of surgical interference is an early and exact localizing diagnosis.

3. Symptomatic Therapy.

This form of therapy may be used with good effect in all diseases in which severe degenerative processes have not gone on.

Degenerative diseases of the cell-fibre-muscle, especially if they be of central origin, give a bad prognosis. Individual symptoms require individual treatment and in this way they may be transitorily dissipated or improved

(a) *Psychical Treatment.*

Quiet consolation, reasonable explanation, prompt dissipation of trifling worries, and assurance in a proper way and manner, and possibly also a certain amount of didactic severity, are often of great and righteous influence.

It is of paramount importance that the physician counteract depressing influences. This is facilitated by a careful and exact examination and a cautious making of the diagnosis.

An exact and careful examination is often of the greatest benefit to the patient on account of the psychical impression that it makes; indeed, it may be said that it is a therapeutic agent of the first order. Hypnotic treatment is only to be thought of when the above procedures are not effective, as the results of this form of treatment are doubtful, frequently of no permanent benefit, and occasionally detrimental.

The necessity of limiting the right to produce hypnosis to the physician should be especially emphasized in this connection. See further under Hysteria.

(b) *Physical Treatment.*

Regulation of nutrition, sleep, and the obtention of quiet, health-conducive surroundings are of fundamental importance. One should seek to obtain these wherever it is possible without any artificial aid. Limitation, or better abstinence, from alcohol, it matters not what lay friends, apothecaries, or even physicians may say about the efficacy of wine to "strengthen the nerves," is one of the first things to be done.

Fruit, cocoa, tea, café au lait, and eventually light wines are to be recommended as nutritive measures.

Hydrotherapy. Cool, not too cold sponging, douching, cool (75 to 85° F.) and warm (90 to 100° F.) baths are most serviceable. Hot packs are often used. After warm baths there should follow a cool rubbing off.

Sojourning in watering-places, especially small

ones (at the seashore or the mountains), is often attended by surprisingly beneficial and rapid results. Massage, gymnastics, and occasionally the reassuring statement of the physician are often of great benefit.

Electricity, whether it operates directly on the somatic functions or indirectly through the mind, is often an ultimate refuge.

The galvanic and faradic currents are the most used. The galvanic current reaches the deeper parts and generally is more serviceable than the faradic. Galvanic treatment is more applicable in painful and deeply situated disease processes (the anodal pole is soothing, the cathodal exciting, therefore the anode should be placed on the painful parts), while the faradic is more used for peripheral diseases, especially for paralysis of the muscles.

Care should be taken not to use too strong a current and never without measuring the current with a galvanometer.

(c) *The Administration of Drugs.*

These constitute, where they are not indicated by some special reason, not the first but the last aid which we invoke.

The chief maxim is here *nil nocere*, for as it is often impossible to help, the risk is unequal. Prevent the unnecessary expenditure of money!

A few medicines operate with considerable certainty on the nutrition: arsenic (acid. arsen. $\frac{1}{60}$ to $\frac{1}{15}$ of a grain, Fowler's solution 5 to 15 drops per diem); iron (Blaud's pills, pills of iron and quinine, and pills of iron and arsenic).

As quieting measures the bromine salts are to be

recommended (bromide of potassium, bromide of sodium, bromide of ammonium), also a mixture of these (1 to 4 drachms per diem), also antipyrin (15 to 30 grains), etc.

As pain-relieving measures the most important are opium, morphine (only with greatest caution, if at all, in the neuroses), and codeine.

As anti-neuralgics: antipyrin (15 to 30 grains), quinine (8 to 20 grains), salicylate of soda (1 to 3 drachms), salicylic acid (30 to 60 grains), phenacetin (8 to 20 grains), antinervin (8 to 20 grains), salophen (8 to 20 grains), analgene (15 to 30 grains).

As hypnotic aids, sulfonal (15 to 30 grains), hypnol (8 to 30 grains), trional (8 to 20 grains), chloral (15 to 30 grains).

In addition to these there are a great many drugs in extensive use purely on theoretical grounds, such as ergotin ($\frac{1}{2}$ to 5 grains), nitrate of silver ($\frac{1}{24}$ to $\frac{1}{6}$ of a grain), strychnine ($\frac{1}{100}$ to $\frac{1}{30}$ grain), atropine ($\frac{1}{120}$ to $\frac{1}{80}$ of a grain), and others. They are in large measure very dangerous poisons and when administered therapeutically have very questionable effects.

Inunctions and counter-irritation are often salutary in their effect.

SECTION V.

Special Pathology and Therapy.

(Plates 69 to 78 and the preceding.)

I. DISEASES OF THE MEMBRANES AND BLOOD-VESSELS OF THE BRAIN.

SUCH diseases injure at first the convexities of the brain, then the adjacent cortical substance. If the seat of the lesion be the base of the brain the cranial nerves become involved, while if it be the spinal cord the spinal roots suffer. Thus result the cardinal nerve symptoms.

1. PACHYMENINGITIS INTERNA HÆMORRHAGICA (HÆMATAOMA OF THE DURA MATER).

Nature.—It consists of an inflammatory, membranous accumulation on the inner surface of the dura with (secondary?) interstitial hemorrhages. Its etiology is unknown.

The disease occurs in adults following alcoholism, in senility, and in dementia paralytica; it also occurs spontaneously.

Course.—The symptoms that it causes (which bespeak the hemorrhagic condition) are headache, symptoms of cortical irritation and paralysis, such as convulsive attacks on one side of the body, and hemiplegia and fever.

Diagnosis.—It must be differentiated from tumor

of the brain (uniformly progressive), hemorrhage into the brain (symptoms of irritation), rarely meningitis (different course, and symptoms referable to the base), and from uræmic coma (examination of the urine and course).

Treatment.—Quiet, ice applications, venesection, calomel, drastics; and to control the symptoms hypnotics and morphine.

2. ACUTE LEPTOMENINGITIS. MENINGITIS OF THE CONVEXITIES.

(a) *Epidemic Cerebro-Spinal Meninigitis.*
(Plate 53, 1.)

Nature.—An epidemic and also sporadic form of infectious disease, the cause of which is the diplococcus of Fraenkel. Its presence (the entrance being through the nose, wounds, and the lymph-blood tract?) and development cause an inflammation of the membranes of the brain and spinal cord. The pathological product consists first of a serous, later of a purulent exudate in the meshes of the pia and especially into the convexities of the hemispheres.

Course.—The disease begins abruptly, most frequently in the young, with a general feeling of malaise, rapidly increasing headache, vomiting of cerebral character, and chills. The disease rapidly becomes severe. Herpes labialis, continuous high fever, severe leucocytosis, stiffness of the neck, and pain in the neck on attempts at bending (irritation symptoms from the upper cervical roots), benumbing of consciousness, all go to corroborate the diagnosis. Then follow delirium, symptoms of cortical irritation, such as convulsions, paralysis of the monoplegic type, and hyperæsthesia. Later still deep coma, trismus, opisthotonus, retention of urine and fæces, and Cheyne-Stokes respiration.

In severe cases death occurs at the end of the first week (meningitis siderans); in severe, protracted cases the disease may last eight weeks and longer. Nevertheless recovery from the severe forms is possible.

The sequelæ of the disease are deafness, lesion of the auditory nerve, blindness, persistent cephalalgia, paralysis, and abscess of the brain.

Diagnosis.—It must be differentiated from typhoid fever (by the slow onset, no herpes, no leucocytosis), from pneumonia (sputum and lung symptoms), from pyæmia (purulent foci, no rigidity of neck), from tubercular meningitis (basal symptoms and secondary to tuberculosis), purulent meningitis (purulent area and not epidemic).

Treatment.—Quiet; wrapping in cold cloths, bathing being painful; ice bags; letting of blood over the mastoid processes; calomel; fly blister to the back of the head; inunctions of mercury; antipyretics and narcotics.

(b) *Purulent Meningitis.*

Nature.—It occurs oftenest with purulent process in the ear (otitis media), caries of the petrous portion of the temporal bone, erysipelas, pyæmia, and other acute infectious diseases. The exciting causes are the different bacteria, streptococcus pyogenes, streptococcus erysipelatis, staphylococcus aureus and albus.

The anatomical development of the disease is as in (*a*), the greatest purulent areas being not only on the convexities but in other locations as well.

Course.—Beginning and course as in (*a*). Paralysis of the cranial nerves, optic neuritis, and pupillary changes occur more frequently than in (*a*).

Diagnosis.—Diagnosis as in (*a*). The origin of the disease should be assiduously sought for.

Treatment as in (*a*). Surgical interference.

3. TUBERCULAR MENINGITIS. BASILAR MENINGITIS.

(Plate 53, 2.)

Nature.—It follows secondarily to primary tuberculosis of the lungs, lymph glands, or bones, and occurs with especial frequency in children. The itinerary of the tubercle bacilli is through the blood-vessels to the pia of the brain, there their activity causes first a fibrinous exudation into the sulci, particularly at the base of the brain, where miliary tubercles are found later in great numbers. When such tubercles are circumscribed in the substance of the brain they cause a tumor formation to which the name of solitary tubercle is given.

Course.—The disease begins gradually, with restlessness, headache, and vomiting. After a time there is an increase in the symptoms of cortical irritation and delirium; convulsions (hydrocephalic cry), trismus, and irregularity of the heart occur.

Somnolency, increase of temperature to $39°$ C., irregular fever, stiffness of the neck, and coma follow.

Paralysis of individual cranial nerves (motor oculi, facial, abducens), monoplegias, aphasia, and death complete the picture.

Diagnosis.—It is to be differentiated from purulent meningitis (which see).

Corroboration of the diagnosis should be sought by examination of the sputum and of fluid obtained by puncture through the lumbar vertebræ for tubercle bacilli.

Treatment as in 2.

4. Syphilitic Meningitis, Gummatous Meningitis, and Brain Syphilis.

(Plate 55, 1).

Nature.—They occur during the secondary and tertiary stages of syphilis, at the earliest a few months after infection, ordinarily the result of characteristic syphilitic new formations (gummatous formation, infiltration, cheesy degeneration) in the pia, especially at the base of the brain, with especial predilection for the blood-vessels which course there. Brain syphilis is described under three varieties:

(*a*) Diffuse syphilitic basal meningitis (cortical infiltration, also localized focal areas). The symptoms are essentially those of basal tubercular meningitis.

(*b*) Isolated gummatous formation in the meninges, rarely in the substance of the brain. The symptoms are those of brain tumor.

(*c*) Characteristic changes in the blood-vessels (endarteritis luetica), with or without diffuse or localized gummatous meningitis. The result of this is to produce narrowing of the lumen of the affected vessels, the encroachment being the consequence of intima proliferation, the result being that autochthonous thrombus formation follows easily and frequently, likewise necrosis. (See Embolism.)

These individual forms may also occur together and therefore the symptoms of brain syphilis are very variable.

The intensity of the symptoms varies very much, the disease being characterized by exacerbations and remissions.

Course.—In the beginning the symptoms are meningitic (headache, vomiting), then follow attacks of loss of consciousness, paralysis of the extremities, convulsions, dementia, apathy associated with lesion of the basal cranial nerves, optic atrophy, optic neuritis, oculomotor and facial palsy. Polyuria and polydipsia are of frequent occurrence.

Hemiplegia occurs very often and is the result of thrombotic softening. Aphasia and epileptiform attacks complete the clinical picture.

Recovery is possible, but some paralysis often remains as a residuum.

Diagnosis.—For the diagnosis, knowledge of a previous specific infection (abortion, cicatrices, infiltrated glands) is of greatest importance, but it is possible and just to make a diagnosis sometimes when such information is denied us.

When an attack of meningitis occurs for which a cause cannot be postulated, one should always think of lues.

Treatment.—Mercurial inunctions, potassium iodide (see General Portion), and symptomatic treatment.

5. Thrombosis of the Sinuses.

Nature.—Coagulation of the blood in the venous sinuses follows secondarily to other neighboring localized areas of disease, especially if they be inflammatory in nature. It is most frequently an extension of the inflammatory process (phlebitis) to the walls of the sinus. Thus it occurs with caries, osteomyelitis of the petrous bone and the mastoid process; with purulent meningitis, abscess of the brain, phlegmon, etc. It also occurs spontaneously in senility, thrombosis of marasmic origin, chlorosis, and cachexia.

The most common location for marasmic thrombus is the superior longitudinal and the transverse sinus. Following disease of the ear the transverse sinus is most frequently affected.

When the origin is purulent the thrombosis results secondarily to infection and the result is metastatic pyæmia (through the jugular vein—right heart, in the lungs).

Symptoms.—The symptoms vary according to the seat of the lesion. Symptoms of meningitis, such as headache, vomiting, convulsions, and coma, are generally the first, while subsequently paralyses and stiffness of the neck occur. When the transverse sinus is thrombosed, there may be œdema over the mastoid process, distention of the jugular vein of the affected side, slowness of circulation through the external jugular vein. When the thrombosis is in the sinus cavernosus, there will be congestion of the ophthalmic veins, chemosis, protrusion of the eyeball, and paralysis of ocular muscles. Thrombosis of the superior longitudinal sinus is generally accompanied by congestion of the veins of the nose (epistaxis).

Diagnosis.—The diagnosis cannot always be made with certainty, as the symptoms are often masked by the original disease (abscess of the brain, meningitis, etc.). The above-described manifestations of congestion go far to substantiate the diagnosis.

Treatment.—Eventually surgical, opening and tamponading; in the mean time symptomatic.

For diseases of the arteries and their consequences see Hemorrhage, Embolism, and Aneurism-formation in the next section.

II. DISEASES OF THE BRAIN SUBSTANCE.

A. ORGANIC DISEASES.

1. DISTURBANCES OF CIRCULATION AND ITS CONSEQUENCES.

(*a*) *Anæmia and Hyperæmia of the Brain.*

Both of these conditions occur as transitory or persistent individual symptoms of different diseases, such as conditions of exhaustion, in fever, and with anæmia of all forms.

Transitory anæmia of the brain, especially of the cortex, causes attacks of syncope; persistent anæmia, after loss of blood, and of previous former anæmia, causes general malaise, somnolence, tired feeling, noises in the ears, vertigo, vomiting, and gaping.

Treatment by infusion of salt solution, horizontal position, clysters, counter-irritation to the skin in the acute and robust. Increase the nutrition by the administration of eggs and milk diet, and by the exhibition of iron and arsenic. In chronic anæmia symptomatic treatment. The administration of alcohol, red wine, is entirely unnecessary and often producive of harm in the treatment of anæmia.

Hyperæmic conditions occur in persons of the apoplectic habit, in plethora, and are accompanied by feelings of heat, cardiac palpitation, vertigo, numb sensations in the head, etc. Arteriosclerosis plays an important rôle in the etiology of this condition.

Treatment.—Regulation of the mode of life, of the taking of alcohol, of the use of tobacco. Baths, sea bathing, exercise, and attention to the heart and kidneys are of first importance.

(b) *Hemorrhage into the Brain.*
(Plates 59, 1 and 2; 60, 1.)

Nature.—The predilection of the arteries of the brain to the formation of miliary and larger aneurisms and the rupture of these aneurismal walls lead to various injuries of the brain substance. The bursting of a small aneurism may be followed by the most severe consequences when adjacent tracts or centres are involved or destroyed.

Such miliary aneurisms may form in any of the arteries of the brain, but they are found most commonly in the branches of the artery of the Sylvian fossa, the most frequent branch to be affected being the lenticulo-striate artery (see Fig. 3, p. 22).

The adjacent brain substance is destroyed to a greater or less extent by the extravasation of blood into it. Following the formation of coagulation or thrombus the blood undergoes cystic degeneration and there is resorption of the exudation with a partial connective-tissue transformation (see General Part). On the surface there remain as sequelæ a defect and in the interior of the brain a cyst.

The most prominent etiological factors leading to such aneurism-formation are arteriosclerosis, the result of chronic alcoholism, syphilis, gout, chronic nephritis, and other unknown conditions.

The most important localization of a hemorrhage is in the vicinity of the cerebral ganglia and the adjacent medullated portions, the internal capsule. This area is supplied with blood, especially by the above-mentioned artery of the fossa of Sylvius, and on account of the frequency with which it is the seat of hemorrhage, it has been called by Charcot the *artère d'hémorrhagie.*

Course.—The hemorrhage is attended by the symptoms of cerebral apoplexy, a shock, a stroke. After the symptoms attending the occurrence of the hemorrhage have ceased, paralysis, hemiplegia, remains. The occurrence of an apoplectic stroke is preceded by certain prodromal symptoms, such as disturbances of circulation manifested by headache, vertigo, noises in the ears, and moderate weakness.

Frequently the attack comes on with lighting-like rapidity without any premonition and without any exciting cause, or with some such exciting cause as bodily strain. The patient feels badly, is dizzy, confused, and quickly becomes unconscious. This constitutes the shock. It may last from a few minutes to several days, depending upon the severity and extent of the lesion. The shock renders the patient unconscious and reactionless, the breathing is slowed, stertorous, often suspended; the pupillary reflex frequently irregular; the extremities are immovable and when one is raised it falls as if inanimate, but occasionally there may be slight spastic manifestations. Retention of urine and fæces develops and oftentimes there is conjugate deviation of the eyeballs. The temperature varies: it may be normal, subnormal, or just preceding the end extremely high.

If death does not follow the shock (depending upon the size and location of the hemorrhage), the patient gradually recovers consciousness, and frequently it is not until consciousness has been restored that the amount of paralysis can be determined.

The paralytic manifestations directly after the shock are very much more extensive than they are after recovery of consciousness or at the end of a week. Those symptoms which disappear during the first week after the onset of the disease are known as indirect focal symptoms. They depend upon the remote activities of the focus through pressure on neighboring portions and they disappear simultaneously with the recovery of this segment of the brain.

The direct focal symptoms, the result of destruction of certain tracts and centres, are on the contrary irreparable and continuous.

The direct, continuous symptoms bespeak the localization and extent of the lesion in very different ways, varying from the transient hemiparesis to severe total hemiplegia with hemianæsthesia and hemianopsia. Apoplexy of the left cerebral hemisphere is naturally often associated with aphasia in its different forms. Apoplexy into the medullary substance may occur without symptoms, while if it be in the cerebral peduncles the symptom complex of alternating hemiplegia may be the result (see Section IV., 3).

The implicated muscles manifest a spastic paralysis with a tendency to the occurrence of contracture formation without degenerative atrophy.* The tendon reflexes are increased, and the skin reflex of the paralyzed side is frequently lost.

On the hemiplegic side there may develop choreic, athetoid, irritation movements (post-hemiplegic hemichorea, hemiathetosis).

In extensive and widespread hemorrhage or more frequently in recurring hemorrhage, the intelligence of the patient suffers markedly, there is pathologically increased irritability and also complete secondary dementia.

Diagnosis.—It is to be differentiated from the coma of uræmia (examination of urine), from diabetic coma (odor of acetone, sugar), from extra-meningeal hemorrhage coming from the middle meningeal artery in fracture of the skull (the history, typical course of varying brain pressure), from embolism of the brain (youth, disease of heart, light attack), and from pachymeningitis (which see).

Topical diagnosis according to paragraph IV., 3, I.

Treatment.—Prophylactic, see Hyperæmia.

At the time of attack, ice bags, quiet, attention to bladder and bowels, venesection, counter-irritation.

After the attack, interdict excitement of all kinds, allow moderate gymnastic practice, electricity, mas-

---

* In very rare instances atrophy of recent onset and rapid course is seen (thalamus lesion?), cerebral atrophy. The hemiplegic gait (dragging) is mentioned on p. 115.

sage, salt baths, mud baths, and symptomatic treatment.

(c) *Embolism of the Brain.*
(Plate 56, 2.)

Nature.—Embolism arises from the formation of a fibrinous coagulation in the interior of an artery which blocks up the calibre of some of the small branches. The embolic material comes mostly from the left side of the heart (mitral endocarditis and from the aorta). It passes through the internal carotid and first becomes lodged in one of the cerebral arteries when it reaches one of sufficiently small calibre, which is most frequently one of the branches of the artery of the fossa of Sylvius and oftenest of the left side. As the cerebral arteries are terminal arteries it follows that there will be no compensatory circulation in the regions to which the blocked-up artery furnishes blood, and the result is that necrosis results through a process of softening (see General Part), the contents of the area of softening are resorbed, and there remains on the surface of the brain a defect, a flat, yellow surface, and in the brain substance a cyst to mark the seat of the lesion.

Obstruction of an artery and its consequences may arise without formation of an embolic process through autochthonus thrombus formation in the arteries of the brain (syphilitic endarteritis obliterans, arteriosclerosis). The areas of softening that result are frequently multiple.

Course.—The clinical manifestations arising from blocking up of an artery of the brain vary but slightly from those of hemorrhage.

Embolism causes an apoplectic stroke as does hemorrhage, but the subsequent coma is not as a rule of such profundity or duration.

The premonitory symptoms of embolism or of autochthonus thrombosis may be very numerous and very severe.

In these there are also direct and indirect focal symptoms. The indirect disappear more rapidly. Those that do not disappear after the first few days are direct symptoms and irreparable.

Depending upon the seat of the lesion there may be more or less transitory hemiparesis, more or less permanent and extensive hemiplegia, aphasia more frequently than with hemorrhage into the brain, alternating paralysis, hemianopsia, etc.

Diagnosis.—For diagnosis see Hemorrhage.

Treatment as in hemorrhage.

(d) *Aneurismal Formation.*
(Plate 54, 1.)

Large aneurisms have their seat in arteries of the base most frequently. Through compression of the substance of the brain, the outgoing nerves, and the pyramidal tracts in the pons and medulla, alternating hemiplegia arises.

In aneurism of the internal carotid there is frequently pulsating exophthalmus.

(e) *Arteriosclerosis of the Brain Arteries.*

This causes, through its disposition to hemorrhage, thrombosis (softening), and aneurismal formation, a group of symptoms such as have been mentioned under hyperæmia of the brain; slight hemiparesis, transitory attacks of vertigo, disturbance of vision, etc., which are dependent upon circulatory disturbance of different degrees.

For treatment see Hyperæmia.

2. INFLAMMATORY DISEASES OF THE BRAIN SUBSTANCE.

(a) *Abscess of the Brain.*

Nature.—A focal collection of pus within the brain substance occurs in consequence of traumatism to the skull or inflammatory processes such as purulent meningitis, caries, otitis media, pyæmia, and abscess of the lungs. Such abscesses as are not attributable to some primary disease are called "idiopathic."

The seat of abscess is most frequently the temporal lobe and the cerebellum, especially secondary to diseases of the ear.

The exciting factors of abscess are the staphylococcus aureus, the diplococcus of Fränkel, the streptococcus pyogenes, the microbe of aphthæ, and the tubercle bacillus.

Course.—The prolonged period of latency which may be up to ten years is characteristic for abscess of the brain. A purulent focus which has been encapsulated may suddenly give rise, sometimes without direct cause, to the gravest focal symptoms, such as perforation into the ventricle or on to the surface.

The symptoms of cerebral abscess are partly paralytic and partly irritative. Constant headache, vomiting, vertigo, irregular fever (sometimes absent), comatose conditions, convulsions, and various focal symptoms, such as hemiplegia, hemianopsia, epileptiform attacks, and coma, may be present. In cerebellar abscess there will be in addition a more or less characteristic disturbance of gait, cerebellar ataxia.

When perforation threatens (extra-meningeal) the symptoms intensify.

Not very rarely decided cachexia and apathy exist.

Diagnosis.—To be excluded are tumor of the brain (nearly always choked disc, no fever). Evidence of a primary purulent focus is of decisive significance and previous trauma is in favor of abscess.

Treatment.—Only surgical, trephining, exploratory puncture, evacuation.

(b) *Acute Non-Purulent Encephalitis.*

Nature.—There are various processes included under this heading which are not well understood.

(α) The hemorrhagic form in which small inflammatory hemorrhagic foci occur accompanied by acute inflammatory symptoms, the foci being found mostly in the cortex, the basal and central ganglia, and of infectious (?), toxic (?) origin.

Course.—The symptoms are made up of meningeal and focal manifestations which latter vary according to the seat of the focus. The prognosis is not always unfavorable.

(β) A similar form leads (in chronic alcoholism) to numerous small hemorrhages into the brain stem, especially in the vicinity of the aqueduct beneath the corpora quadrigemina. It produces ocular palsies, acute ophthalmoplegia.

(γ) Here belongs in part the as yet not very well known process on which are dependent the infantile cerebral palsies (Plate 54, 2). Infantile palsy occurs as congenital or in the first months of life and is dependent upon different disease processes.

Course.—In the first months of life the child takes sick suddenly with grave meningitic symptoms, which, however, soon subside. These symptoms are caused by inflammatory processes in the central convolutions which leave as a residuum, after the subsidence of the inflammation, scars, defects (porencephaly) in the diseased convolutions.

Therefore we find as clinical symptoms after cessation of the initiatory symptoms palsies of the extremities, spastic palsies of course (when the disease is double-sided, spastic paraplegia). The growth of the paralyzed extremities, an arm, a leg, or the whole side, does not keep pace with the unaffected one. The whole half of the body may be atrophic (spastic hemiatrophy).

Symptoms of irritation, such as epileptiform attacks, caused by irritation from the cortical scar and athetosis, are of frequent occurrence.

Treatment.—Symptomatic. See Meningitis.

3. TUMOR OF THE BRAIN.
(Plate 60, 2.)

Nature.—Tumors may develop within the skull secondarily to trauma; they may be metastatic from other tumors in distant part of the body, or they may be spontaneous.

They may spring

From the bones (sarcoma, carcinoma, osteoma, gumma);

From the membranes (sarcoma, carcinoma, fibroma, solitary tubercle, gumma);

From the blood-vessels (gumma, aneurism); and from the substance of the brain (glioma, sarcoma, carcinoma, metastatic and primary, and cysticercus).

The most common are glioma, gumma, and sarcoma in the adult, and solitary tubercle in children.

The symptoms vary according to the size of the tumor, the rapidity of its growth, and with the multiplicity and seat of the growth.

All tumors have a series of symptoms which are dependent upon the presence of an increasing foreign body within the closed skull space; these may be called the general tumor symptoms. The other symptoms or special symptoms depend upon the seat and variety of the new growth.

Course.—The general symptoms of brain tumor are the result of space encroachment in the cranial cavity and the resulting difference in intracranial pressure, which acts in a sinister way not only upon the brain substance but upon the lymph and blood circulation.

These symptoms are:

(*a*) Headache, continuous, diffuse, dull, and intense.

(*b*) Cerebral vomiting not associated with the taking of food and occurring periodically.

(*c*) Vertigo, stupidity, drowsiness, hebetude.

(*d*) Bradycardia.

(*e*) Apoplectiform and epileptiform attacks.

(*f*) Choked discs, eventually optic neuritis (Part IV., Section III.) are the cardinal symptoms of brain tumor.

The focal symptoms, the directly operative and the remote, vary with the seat of the tumor and consist of monoplegia, hemiplegia, eventually crossed cranial nerve palsies, hemianæsthesia, hemianopsia, cerebellar ataxia (involvement of corpora quadrigemina, cerebellum), aphasia, and basal symptoms (see Focal Diseases, Part IV., 3).

Focal symptoms may be absent, only suspected.

Tumor of the corpora quadrigemina may produce acute internal hydrocephalus by causing pressure on the large veins of Galen, which prevent the flow into the third ventricles.

Diagnosis.—The diagnosis is first made from the general symptomatology.

To be excluded are, abscess of the brain (which see), hydrocephalus (see also). The seat of the neoplasm is to be determined according to the focal symptoms. See Section IV., 3, for the localizing symptoms referable to the anterior, middle, and posterior fossæ of the skull, and for the individual lobes of the brain, the basal ganglia, the cerebellum, etc. For the topical diagnosis only the continuous, progressive, direct focal symptoms are considered, not the indirect and transitory which vary in their intensity. Nevertheless it is not always possible to separate one from the other. The limit and the kind of the tumor are then to be taken into consideration (whether metastatic, tuberculous, luetic, etc.).

Pathological.—For many tumors the location is characteristic. Tumors of the base are likely to be either gumma or sarcoma. Of the brain substance, glioma; cerebellar-pons tumor, solitary tubercle, glioma. Sarcomata are of rapid, gliomata of slow development.

Treatment.—On the slightest suspicion inunctions

of mercury and iodide of potassium. When the tumor is cortical, surgical aid may be invoked (see General Therapy).

Aside from this, symptomatic treatment, narcotics, bromides.

4. INTERNAL HYDROCEPHALUS.

Nature.—By this term is understood a pathological increase of the amount of the cerebro-spinal fluid in the ventricular spaces (lateral and third ventricles).

External hydrocephalus is a collection of fluid in the subarachnoidal space which communicates with the ventricles. Internal hydrocephalus may be congenital or acquired. As causative are postulated inflammatory processes of the ependyma of the ventricles and the choroid plexus, and in part an obstruction to the escape of the fluid from the third ventricle, the result of inflammatory changes.

If the hydrocephalus reaches a high degree, it produces flattening of the cerebral convolutions, and distinct decrease of the brain substance results from pressure atrophy. The circumference of the skull increases up to 60 cm., the normal being under 50 for children. Accumulation of fluid in the third ventricle causes an outward convexity of the walls of the ventricles and pressure upon the surrounding structures or parts, such as the optic nerves, the chiasm (bitemporal hemianopsia), the motor oculi and facial nerves.

Course.—In congenital hydrocephalus, as well as in hydrocephalus that develops in early childhood, the occurrence of the disease causes idiocy, double-sided spastic-paralytic phenomena, and epileptic attacks. These children soon die.

When the disease develops later in life, in adults, symptoms of meningitis, such as headache, vomiting, coma, optic neuritis, absence of fever, attended by exacerbations and remissions, mark the development of the hydrocephalus. Choked discs may also be present.

Diagnosis.—Tumor of the brain cannot always be excluded with certainty, but absence of increase in circumference of the skull, rapid development of the tumor symptoms, and absence of remissions speak in favor of the latter.

Treatment.—Surgical procedures, puncture of the ventricle, or of the central canal of the spinal cord in the lumbar region, are of doubtful utility. Counter-irritation may be made to the shaved scalp. Symptomatic treatment.

5. DEMENTIA PARALYTICA, PROGRESSIVE PARALYSIS.

Nature.—Progressive paralysis is dependent upon a degenerative process in the brain. In the cortex especially numerous non-medullated and medullated fibres as well as cells are primarily destroyed. The change is most marked in the frontal lobes. As the result of destruction of the fibres there is diminution in the size of the cortex and eventually atrophy of the entire brain.

Aside from this there are found chronic changes of an inflammatory nature in the blood-vessels, in the glia, and in the meninges; hydrocephalus, etc.

The disease occurs more frequently in males, and with striking frequency in those who have had previous luetic affection, so that, like tabes, it is recognized as metasyphilitic. It also develops after trauma.

Course.— The degeneration of the cortical nerve elements leads to a progressive diminution of psychical functions, of the intelligence, character, memory, speech, and especially of the faculty of association.

Change in the character of the patient is frequently the earliest symptom; this may manifest itself by stupidity, irritability, tendency to excesses, inability to do mental labor, and by immorality.

If the patient now consults a physician the following characteristic symptoms are to be observed:

(1) Immobility of the pupils, unilateral or bilateral asymmetry of the pupils.

(2) Paralytic speech disturbances, literal paraphasia, negligent, incorrect speech, omission and displacement of letters and syllables (for other speech disturbances see IV., 4).

Analogous disturbances are to be seen in the patient's writing and reading, omission of letters, syllables, words, and punctuation.

(3) Loss of patellar reflex. Dementia paralytica may be associated with symptoms of tabes and also with symptoms pointing to involvement of the lateral columns of the cord (increase of reflexes).

(4) Tremor of the hands, the tongue, and tremulousness of the lips on speaking.

As the disease progresses paralytic attacks in the shape of apoplectiform loss of consciousness appear, and these may or may not be accompanied by epileptiform attacks. After an attack is over the symptoms of paralysis disappear.

The psychical shortcomings increase and soon expansive ideas, feelings of exaltation, or attacks of frenzy occur. The course of the disease is marked by periods of remission, during which time the patient may for a considerable period appear quite normal, but sooner or later the old conditions show themselves and unfortunately in a more aggravated form.

Toward the end there is complete loss of all the psychical faculties, complete dementia.

Death follows from exhaustion, or from pneumonia caused by the entrance of foreign bodies into the larynx, from a few months to two or three years after its onset.

Diagnosis.—It is to be differentiated from severe neurasthenia (continued absence of objective symptoms), from multiple sclerosis (which see), and from syphilis of the brain (well-marked focal symptoms).

Treatment.—Where it is possible, treatment is best carried out in an institution. Symptomatic treatment. Anti-luetic treatment is rather harmful.

6. OPHTHALMOPLEGIA.

Double-sided paralysis of the eye muscles from disease of the nuclei of this nerve beneath the corpora quadrigemina arises with many different diseases.

(a) Acute Ophthalmoplegia.

This may be dependent upon the above-described hemorrhagic encephalitis, upon other hemorrhagic embolic processes, and it may develop with tumors.

(b) Chronic Progressive Ophthalmoplegia.
(Plate 62, 3.)

This, with progressive bulbar palsy about to be described, belongs to the system diseases of the neuron cells (see Section IV., 3).

There occurs a slow, widespread cell atrophy in the motor oculi nucleus, which in the course of years leads to a total paralysis of the external muscles (more rarely the internal) of the eyeball. Naturally with such a process the fibres of the motor oculi likewise atrophy.

This disease may occur by itself or it may be a part of the symptom complex of tabes, dementia paralytica, bulbar paralysis, and multiple sclerosis.

Treatment.—Possibly anti-luetic.

7. BULBAR PARALYSIS.

(a) An acute form may be the result of a thrombotic, embolic, hemorrhagic, or inflammatory process.

The most common introductory apoplectiform symptoms are paralysis of the tongue and throat; affection of the facial, abducens, and trigeminal nerves, and disturbance of phonation and of breathing. Bulbar speech, paralysis of the extremities, and sudden death from paralysis of the heart and respiration.

Treatment.—Possibly anti-luetic.

(*b*) Chronic progressive bulbar paralysis (Plates 72, 1; 56, 1). (Glosso-labio-laryngeal paralysis progressiva.)

Nature.—This is dependent upon a slow progressive disappearance of the cells in the nuclear region of the medulla, especially in the nuclei of the motor nerves (peripheral neuron cells), the hypoglossus, the facial, the vagus-accessorius, and the trigeminus. Aside from this there is often disease of the pyramidal tracts and the anterior horn cells of the spinal cord and there may be every gradation from simple bulbar paralysis to amyotrophic lateral sclerosis with bulbar symptoms (see further under "System Diseases," IV., 3).

Concerning the actual cause of this disease but little is really known.

Course.—The clinical manifestations come on very slowly. The first symptom is disturbance in speech of a bulbar type; the letters *r, s, l* are the first to be formed defectively. After this there develops double-sided atrophy (with reaction of degeneration) and weakness of the tongue and lips which is associated with fibrillary twitchings and weakness of the power of swallowing, of phonation, and less frequently of the muscles of mastication. Eventually entire loss of expression of the features (only the inferior portion of the face is affected), salivation, paralysis of swallowing and of speech (anarthria) follow. Disturbances of sensibility are wanting.

There are often psychical symptoms such as disposition to cry, and if other systems are affected, such as the pyramids, there will be spastic paralysis of the extremities. Atrophy of the muscles of the hands (see Amyotrophic lateral sclerosis) may also occur.

Death occurs from exhaustion and from the pneumonia of swallowing.

Treatment.—Symptomatic, electricity, baths, if necessary feeding by means of stomach tube. Arsenic, strychnine, iodides, and atropine are useless.

It is impossible to replace either the diseased ganglionic cells or the atrophied fibre processes.

Bulbar paralysis may be simulated by other diseases.

(a) Pseudo-bulbar paralysis is caused by a double-sided lesion (embolism or hemorrhage) of the cortical substance of the hemispheres. This may be the condition when the facial-hypoglossal centres are injured on both sides and the symptoms of bulbar paralysis follow, without, however, any attending muscular atrophy. There may be also at the same time diseased areas in the medulla.

(b) Bulbar paralysis without anatomical findings. This is a combination of bulbar symptoms with weakness of the extremities and other muscles, paresis without atrophy. Heretofore in these cases anatomical examination of the nervous system has been negative.

Diagnosis.—These forms cannot be differentiated with certainty from the acute form of bulbar palsy. In hysteria and multiple sclerosis very similar clinical pictures may develop. One must weigh with great care all the individual points and factors in the disease before coming to a diagnosis.

8. DISEASE OF THE CEREBELLUM

has been spoken of under tumor and abscess.

Other diseases of the cerebellum, such as atrophy and sclerosis, are of rare occurrence.

9. MULTIPLE CEREBRO-SPINAL SCLEROSIS.

See Diseases of the spinal cord.

B. BRAIN DISEASES OF UNKNOWN NATURE AND SEAT (NEUROSES OF THE BRAIN).

1. Neurasthenia.

Nature.—This with hysteria and hypochondria predisposes to the severer psychical diseases (psychoses).

Neurasthenia is in part the result of faulty education and modes of life, mental strain, excesses of all kinds, and is in part (especially in those who are handicapped through heredity) the manifestation of a predisposition to weakness and irritability of the psychical and bodily functions.

The patient is in bad spirits, has feelings of anxiety, and is irritable; the will power and executive power are diminished and he is especially deficient in power of concentration.

Other manifestations especially deserving of mention are: easily produced fatigue, morbid self-observation (nosophobia, syphilophobia), sleeplessness, head pressure, tremor, and palpitation of the heart.

A number of other bodily symptoms may complicate the clinical picture; these are abnormal sensations of all kinds, pain, disturbance of the functions of the stomach (hyperacidity, nervous dyspepsia) and intestine, and sexual incapacity. See also Occupation neuroses.

In severe cases there develop delusions, claustrophobia, without, however, any other manifestations of psychical disease. The typical course of neurasthenia is an intermittent one; for a long time the patient feels very well, when suddenly he is thrown back into the former wretched condition.

Diagnosis.—To be excluded by careful and precise examination of the urine, lungs, heart, stomach, are other bodily diseases, dementia paralytica, multiple sclerosis, and hysteria.

Treatment.—Congenial surroundings; careful examination is an important therapeutic factor as well as a diagnostic help. It is a mistake to send such patients away without any objective point in view; the neurasthenic is really sick and should be treated accordingly. The nutrition should be stimulated by food, bathing, massage, gymnastics, electricity, exercise, and by the administration of iron, arsenic, quinine. No morphine, no excessive medication.

2. *Hypochondria.*

Nature.—Patients with this disease are given, like in neurasthenia, to morbid self-observation and examination. Although their complaints change as to localization and intensity, the perverted sensations of hypochondriacs are more constant and localized than those of neurasthenics. Fears concerning the proper performance of the abdominal functions furnish the most common grounds for complaint. The influence of these sensations on the mind is lasting and they are worked over in the patient's thoughts and systematized. The dissonance of the hypochondriac is much more deeply seated than the superficial form in neurasthenia. The neurasthenic speaks openly and freely, the hypochondriac frequently nurses and keeps his complaint to himself.

Diagnosis.—To be excluded are diseases of the stomach and intestine, such as carcinoma and tænia. To differentiate it from paranoia with its systematized delusion it is necessary to observe the course of the disease carefully for some time.

For treatment see neurasthenia.

3. Hysteria.

Nature.—Hysteria is a psychosis, characterized by pathological disturbance of mental conception and the will, which leads to an inexhaustibly large number of functional anomalies of the motor and sensory parts of the body for the explanation of which no organic lesion can be found. In all its manifestations it is psychically tinged.

Course.—It occurs most often in the young. Heredity, faulty education, and alterations of feelings and sentiments are of great importance in the causation of the disease.

The mental condition in hysteria is so changed that the patient becomes more susceptible to strange influences (suggestion) as well as to ideas originating within himself (auto-suggestion), and thus becomes a victim to psychical and bodily reaction. These patients are irritable, moody, capricious, distracted, and explosive. Their intellectual faculties are unaffected.

The bodily symptoms are:

(1) Persistingly existing stigmata, which, however, may be absent.

(2) Hysterical attacks occurring intermittently.

(*1*) *The Stigmata.*

(*a*) Hemianæsthesia, total loss of sensibility of an entire half of the body or of a single area, for all forms of irritation (analgesia, etc.).

At the same time the senses of smell, taste, and hearing may be lost on the same side of the body. Color sensibility may be disturbed (achromatopsia) and the visual fields narrowed. Patient easily fatigued.

The hemianæsthesia is caused by the loss of association of impressions conveyed to consciousness.

(*b*) Hyperæsthesia of a half of the body, or a portion of it, manifested by excessive sensitiveness (ovarian pressure pain, vertebral column pain). Pressure on these points (hysterogenic zones) will often give rise to an hysterical attack (loss of central associative limitation).

Pain of every kind, abnormal sensations, globus hystericus, sensation of a ball in stomach are very variable symptoms.

(*c*) Hysterical paralyses may simulate every form of paralysis. They are of central origin and dependent upon paralysis of the will. They may disappear as abruptly as they came.

The most common forms are hysterical paralysis of the vocal cords (aphonia), paralysis of writing, speaking, and swallowing, hysterical hemiplegia without facial and hypoglossal involvement, hysterical disturbance of gait (inability to stand, astasia, and to go, abasia), and compulsory movements of every variety. Retention of urine.

(*d*) Hysterical contractures are dependent upon

central irritations of a purely functional sort. There may be the different forms of contracture in the flexors and extensors of the head and extremities.

(2) *Hysterical Attacks.*

Irritative symptoms of central origin, which appear in various forms and combinations, partly of spontaneous origin and in part called into being by some reflex excitation such as fright, affectation, sensations, etc.

Clonic, tonic convulsions of the extremities and of the body, quickly changing from one to the other, are the principal symptom. Breathing is increased and dyspnœic, consciousness is seldom altered to any extent, but there is loss of memory for the attack (amnesia).

The convulsions produce in a wild manner the most bizarre positions of the body (clownism, arc de cercle); spasmodic movements of the extremities (grand hysteria), trismus, convulsive movements of swallowing, laughing, crying, etc., all occur with varying severity.

During the attack there are often hallucinations and disagreeable mental possessions, which may be manifested by shrieks, insults, gesticulations, grimaces, etc.

There never occurs such deep coma as in an epileptic attack. The reflexes remain intact. In order to awaken the patient, all that is necessary in the milder attacks is to dowse him with water, while in the more severe attacks the painful prick of a pin will succeed (soles).

Characteristic of hysterical attacks are the exaggerated, mannerless, changeable manifestation of the convulsions, the terminableness of the attack, and the susceptibility of the attack to psychical influence (suggestion).

To the psychical conditions are attributable different disease manifestations, which in some instances are known under individual names, such for example as akinesia algera, pain on movement without any attributable cause therefor; chorea electrica, sudden backward movements of the body; paramyoclonus multiplex, changeable variously localized clonic contraction of muscular fibres of biceps, quadriceps, supinator longus, deltoid, etc., and astasia-abasia, etc.

Diagnosis.—For the diagnosis it is necessary to take into consideration the entire picture which has been delineated above; the presence of individual symptoms is not sufficient for a diagnosis. The significance of stigmata is great. The diagnosis is to be reached only after most careful examination.

It is not possible in every case to differentiate it from neurasthenia, yet in most instances this or that symptom will bespeak the one or the other disease.

In the more severe forms of hysteria the psychical disturbance may be brought prominently into the foreground (moral defects).

Treatment.—Judicious mental discipline, regular and absorbing occupation, awakening of confidence in the will power, exclusion of nervous, over-considerate members of the family (treatment in an institution), considering the attacks as a mere bagatelle, regulation of the nutrition, massage, douches, etc. Possibly the administration of iron and bromides (their utility is insignificant compared with other measures), but never morphine.

Hypnosis affects only individual symptoms, very rarely the disease itself, and oftentimes it causes new symptoms.

The influence of a rational, not directly hypnotic, psychical (suggestive) mode of treatment is by far the better.

4. The Traumatic Neuroses.

Nature.—Some of these cases belong with neurasthenia, others with hypochondria and hysteria. They have in common with these that they result from trauma, psychical and physical, and mostly in individuals who are predisposed to nervous disease by heredity, alcoholism, etc.

The possibility, moreover, that the motive is to obtain money such as from accident insurance companies should always be taken into consideration with these patients. The border line between purposive exaggeration and pathological psychical limitation is often difficult to lay down.

Opinion.—On therapeutic grounds one must determine not so much the injury as how far it incapacitates the patient from work. Only in exceptional cases should one give a definite opinion before he has had abundant opportunity to watch and examine the patient.

In this category belong the neuroses developing after railway accidents (railway spine) and other accidents.

For treatment see neurasthenia and hysteria.

5. Hemicrania, Migraine.

Nature.—This consists of periodical, recurring attacks of severe headache limited to one-half of the head and associated with drowsiness and vomiting. It arises spontaneously or in combination with other nervous symptoms, and it often occurs in families. In some cases it is possibly reflex (from the nose, stomach, genital organs) in origin. The duration and frequency of attacks are very variable.

Disturbances of vision, such as scintillating scotomata, obscuration of part of the field of vision, etc., may accompany or precede the attack. Hyperacidity (toxic manifestation?) is often present.

Treatment.—Carlsbad cure, treatment at bathing establishments, iron, arsenic, removal of any attributable cause. Symptomatic, the salicylates, antipyrin, caffeine, mustard baths, counter-irritation, rest. Never morphine!

6. Cephalalgia, Habitual Headache.

Headache is frequently only a symptom of other disease, such as fever, disturbances of circulation, intoxications, tumor, etc. In nervous patients it may be a cardinal symptom.

It is to be differentiated from the condition just mentioned above, from any diseased condition in the anterior cranial fossa, and from ocular conditions such as myopia, asthenopia, etc.

Treatment.—See neurasthenia. The nervines.

7. Genuine Epilepsy.

By epilepsy is understood the occurrence of periodical attacks of so-called epileptic fits. From the genuine form is to be separated the form known as Jacksonian or symptomatic epilepsy, which has been considered above in connection with focal disease of the brain.

The occurrence of an epileptic attack is supposed to be dependent upon alteration in the brain cortex, especially of the central convolutions; nevertheless the cortical origin is disputed. It is supposed that the cause of an attack is some irritation of the cortex, analogous to what is probably the explanation of attacks of symptomatic epilepsy, where the irritation is from a scar, splinters of bone, or inflammatory products, etc. From this focal point where the irritation is supposed to be, the excitation passes (through association tracts?) to other neighboring motor centres, inhibits or excites them, and the result is an epileptic attack.

As causative of genuine epilepsy may be reckoned heredity, trauma, and the infectious diseases. It occurs most frequently in youth and varies greatly in intensity and duration. The attacks occur either in a severe (epilepsy major, *grand mal*) or mild (epilepsy minor, *petit mal*) form, and they may also be manifested by a psychical equivalent, psychical epilepsy.

Attacks.—A typical attack begins with the aura, headache, confusion, vertigo, optic or acoustic sensations, which lead up to the attack.

The patient falls to the ground unconscious and

lies in a condition of tonic spasm, partly in a flexed tonic condition but mostly in a condition of extensor spasm. The reflexes are lost, the pupils dilated. After a time (up to half a minute) this is followed by a clonic manifestation, consisting of severe, violent convulsions of the muscles of the extremities and other muscles, accompanied by frothing at the mouth and biting of the tongue.

After a few minutes the severity of the spasms abates and the patient passes into a comatose condition, from which he awakens after a longer or shorter time. Headache, mental confusion, fatigue may be of much longer duration. There is generally partial or complete amnesia of the attack. In milder attacks the period of tonic and clonic spasms is very much shorter, while in the equivalent form there may be simply a momentary dream-like confusion (state of confusion), but maniacal excitement, delirium, and hallucinations may also occur. The attacks may come only at night, when the disease is called nocturnal epilepsy. If attacks seem to follow one another closely and without interim, this dangerous condition is known as status epilepticus.

In many cases, but not in all, there occurs after long continuance of the disease, psychical degeneration.

Diagnosis.—To be excluded are: hysterical attacks, Jacksonian epilepsy (focal symptoms when the disease first showed itself), attacks of uræmia (urine), simulation (reflexes not lost, prick of the sole), epileptiform attacks of dementia paralytica (the history and the longer duration).

Treatment.—Bromide of potassium ℨ iss. to ℨ iv.

daily and opium in from 1 grain to 15 grains per day for a long time [6 weeks], followed by bromides, are the most useful measures. Hydrotherapy, abstinence from alcohol, vegetable diet. In well-marked focal attacks the question of surgical interference may arise.

8. Infantile Eclampsia.

Nature.—It occurs in children in the first years of life. A great number of diseases may lead up to the occurrence of convulsions in children, such as affection of the stomach, intestines, teeth, rhachitis, etc. (defective reflex inhibition in the nervous system of the child, the pyramidal tracts are not yet medullated).

The convulsive attacks consist of tonic and clonic convulsions of the entire body, which are frequently accompanied by shrieks of a convulsive nature. Epileptic attacks later in life rarely follow.

Diagnosis.—To be excluded is tuberculous meningitis (basal symptoms, fever).

Treatment.—Remove causative conditions (worms, defects of nutrition), administer bromides, warm baths, counter-irritation, cold sponging.

9. Chorea Minor; Sydenham's Chorea; St. Vitus' Dance.

Nature.—A disease of childhood. Neither its cause nor its location is known. It is thought to be of infectious origin. Of importance is its frequent association with acute rheumatic polyarthritis and rheumatic endocarditis. It also often occurs with pregnancy, this form being especially severe.

Chorea.—Chorea consists of involuntary, uncontrollable, irregular movements of the fingers, hands, arms, legs, head, and face muscles and tongue. In all the muscles submissive to the action of the will there is a condition of continuous activity, so that it is impossible to keep the body or the extremities quiet for a moment. Individual cases vary from a slight involuntary movement of the finger (which is often looked on in the schools as a manifestation of naughtiness) to the most extravagant action of the muscles (*folie musculaire*). In these severe cases speech, walking, and eating are impossible. Chorea may limit itself to one-half of the body (hemichorea).

At night the restlessness and movements may cease. The duration of the disease is from a month to a year. Relapses are frequent. It is exceptional for recovery not to take place eventually.

Treatment.—Arsenic, antipyrin, salol, bathing, improved nutrition, forbid going to school.

Entirely different from chorea minor is

10. Chronic Hereditary Chorea (*Huntington's Chorea*).

It occurs at a more advanced age and in families. The disease is of unknown nature and consists of irritation, motorial symptoms, individual contractions which increase in extent and intensity with years. This is associated with progressive failure of the mental faculties. Recovery is unknown.

Differing from this is the *maladie des tics convulsifs* (*tic général*), insomuch as the development of the involuntary co-ordinated convulsive movements of the face, throat, and hand muscles is associated with compulsorily produced speech movements, profanity, insults, etc. (echolalia, coprolalia). Its course is very chronic.

For chorea electrica see hysteria.

11. *Paralysis Agitans* (*Parkinson's Disease*).

Nature.—This occurs only in the more advanced periods of life. Its cause is unknown. Anxiety, care, etc., have been attributed as causes.

Course.—It begins with a peculiar, slow, rhythmical tremor, movement of the arms and hands (similar to the movements of rolling a pill). This tremor continues during rest and is made worse by excitement. In the course of years the tremor invades all extremities, the head and lips become involved. The posture of the body is characteristic, a forward bent, rigid, stooping attitude. During sleep the movements cease as a rule. Toward the end paralytic manifestations appear. The disease is incurable.

Treatment.—Rest, nutrition, bromides.

12. *Myotonia Congenita* (*Thomsen's Disease*).

This is a rare disease of a family type. The patient is prevented from making any voluntary movement, especially walking, by spontaneously occurring spastic resistance of the muscles. Psychical irritation increases the spasticity of the muscles (similarity to stuttering).

Treatment.—Baths, gymnastics.

III. THE DISEASES OF THE SPINAL CORD.

A. Diseases which are Preponderatingly Focal. Transverse Lesions.

Diseases of the Membranes of the Spinal Cord.

The membranes of the spinal cord become diseased most often secondarily to those of the brain, rarely alone. The most frequent combination is cerebrospinal meningitis and the purulent form; less frequently the tubercular form. The anatomical conditions are similar to those in the same disease of the brain.

The clinical symptoms pointing to involvement of the spinal meninges are often obscured by the preponderance of those pointing to brain involvement. Irritation symptoms pointing to involvement of the spinal roots (reflexly through the posterior, or direct through the anterior) are stiffness of the neck (roots of the upper cervical cord), opisthotonus (dorsal cord roots), radiating pains, and paralysis of the bladder.

Nevertheless these are not so pathognomonic, for frequently cerebral influence cannot be eliminated.

Individual forms are:

1. Hypertrophic Cervical Pachymeningitis.

Nature.—It consists in the formation of a thick inflammatory membrane on the dura, especially on the cervical portion, which adheres to the other membranes and causes compression of the cervical roots and the cervical cord. The etiology of the disease is unknown.

Course.—The symptoms are those of focal lesion in the cervical cord (see Part IV., 3), combined with severe radiating irritative root symptoms, such as severe neuralgic pains in the back of neck and shoulders, paralysis and muscular atrophy in the regions supplied by the ulnar and median nerves; the small muscles of the hand and the flexors of the forearm are especially involved, and as a result of the activity of the antagonistic muscles (the musculo-spiral musculature) the hands assume the position known as the "preacher hands."

Improvement is possible.

Treatment.—Possibly anti-luetic, electricity, symptomatic; actual cautery to the neck.

2. Syphilitic Spinal Meningitis.

Nature.—This is analogous to syphilis of the cerebral meninges, and is attended by the formation of a diffuse membranous exudation. The lesion is in the meninges of the spinal cord (pia mater) either in connection with syphilis of the brain or without it. This syphilitic disease is characterized by the presence of a gummatous, flattened new formation, which macroscopically appears as a thick, resistant, gelatinous covering and which when seen on cross section (see Fig. 13) seems to surround the entire cord and in certain places to grow into the substance of the cord. The most common location is the dorsal cord. The arteries show changes similar to those that are found in endarteritis luetica in the brain. The nerve roots are penetrated by the new growth, and tumor-like gummatous formations are often present. In accordance with these findings the conditions are referred to as syphilitic meningo-myelitis, meningo-neuritis, and meningo-arteritis.

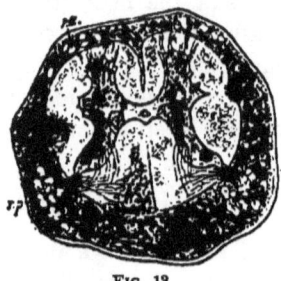

Fig. 13.

Course.—The symptoms, which are the result of compression, vary according to the extent of the lesion and the intensity of the pathological process. They are severe pains occurring paroxysmally, spastic paraparesis of the legs, which is not infrequently more severe in one leg than in its fellow (suggestive of half-sided lesion), increased reflexes, disturbance of the function of the bladder and rectum (retention), mild disturbances of sensibility such as paræsthesia.

Diagnosis.—Especially characteristic is the variable coming and going of the symptoms; improvement and recurrence alternate. Eventually there are frequently brain symptoms such as paralytic phenomena (basal).

The knowledge of previous specific infection is the most important diagnostic factor.

Treatment.—Anti-luetic treatment even in suspicious cases, inunction cure, injection cure, potassium iodide. Frequently we can expect only partial recovery (when the degenerative process is extensive). Symptomatic. Baths.

3. Compression of the Spinal Cord (Compression Myelitis).

(Plates 54, 2 and 3; 69, 3.)

Nature.—Although compression of the spinal cord may be the result of different processes, it is possible to speak of the various conditions here together. Frequently in the diagnosis of individual cases we are able to say that compression exists; but how it is compressed remains in the dark.

Compression of the spinal cord may be caused:

1. By disease of the bone, such as caries of the body of the vertebræ, which allows of cord compres-

sion by purulent foci; proliferation of granulations in the membranes of the cord (meningitis tuberculosa); sequestra; by sinking in of the vertebræ, the result of undermining of the vertebral canal; by carcinoma of the vertebræ and by traumatic fracture and dislocation of the spinal column.

2. By diseases of the membranes, pachymeningitis, syphilitic meningitis, tubercular meningitis, by sarcoma, lipoma, fibroma, and echinococci formations in the membranes.

3. By diseases of the spinal cord, such as glioma, sarcoma, and gumma.

By far the most common causes of compression are caries and carcinoma.

As a result of compression there occurs severe injury to the involved segment of the cord and the nerve cells and fibres that are connected with it. These undergo degeneration and become destroyed, nevertheless an actual myelitis is not the rule.

Course.—The most important symptoms of compression are the irritation root-symptoms, consisting of severe radiating pains which are frequently manifested in the peripheral distribution of the nerves (neuralgiform) and partly girdle-like about the trunk, soon passing into the extremities. These frequently produce reflex contractions in the muscles of the extremities (muscles that do not contract to the will). Hyperæsthesia and paræsthesia may accompany these pains. As an irritation root-symptom must be classified the stiffness of the neck and of the back. The other symptoms, varying according to the seat of the lesion in the cervical, dorsal, lumbar, sacral cord or cauda equina, consist of paralysis, partly spastic, partly flaccid, atrophy of the muscles, and disturbances of sensibility and of the bladder. The most frequent form is compression of the dorsal cord, and in this there are spastic paraplegia of the legs, disturbances of the bladder and of sensibility, the arms remaining intact.

Death occurs from the original disease and from cystitis, exhaustion, and bed-sores.

Diagnosis.—First it is necessary to diagnose a compression, then the level of the compression. To do this one succeeds best in the following way:

One first determines which extremity is paralyzed. When only the legs are affected the patellar reflexes are tested. If these are lost, the seat of the lesion is in the lumbar cord (possibly the roots of the cauda equina). The patellar reflex is preserved when the sciatic fibres alone are compressed. If the knee jerk be increased, the seat of the lesion must be above the reflex arc, that is, in the dorsal or cervical cord.

If there be no disturbance of the arms, especially no muscular atrophy of the hands, the seat of lesion is in the dorsal cord.

More exact localization is possible by examination of the muscular atrophy which eventually occurs (reaction of degeneration), and by the disturbances of sensibility, the height it reaches, etc. The seat of the lesion is often placed too high; we are not yet sufficiently familiar with spinal localization to fix accurately the disease process in any single segment.

When one has localized the seat of the lesion as precisely as possible he should then compare the segment or segments with Plate 23 and determine what spinal process or processes the lesion corresponds to, and then the locality thus decided upon and the mobility of the spinal column of it should be examined.

In order to diagnose the variety of compression process it is necessary to bear in mind the following facts:

In favor of caries speak youth of the patient, change in the spinal column, angular kyphosis, pain on pressure, the occurrence of gravitation abscesses, fever if cystitis can be excluded, tuberculosis of other parts, and a protracted course.

In favor of carcinoma speak a more advanced age, cachexia, metastasis, a more rapid course, and intense pain.

In favor of glioma of the cord are symptoms limited to one side.

In spite of all these considerations it is frequently impossible to reach a positive conclusion and it must be remembered that compression itself, in spite of the proliferation (central gliosis), outspoken symptoms which it generally causes, can-not always be detected.

Treatment.—In caries spinal extension, arsenic, operative (?); in tumor surgical (see General Part); in most cases symptomatic, regulation of the functions of the bladder (catheter), morphine.

4. *Acute and Chronic Myelitis.*
(Plates 69, 1 and 2; 57, 2 and 3.)

Nature.—Under the name myelitis are included a number of diseases which are very unlike from an etiological standpoint and a number of them from an anatomical standpoint as well.

In all of them there is found in certain areas of the

cord a diffuse focus of degeneration or a degeneration of nerve bundles, and in its place a mass of newly formed glia tissue with more or less marked round-cell infiltration, changes in the walls of the blood-vessels, etc.

Such processes are probably mostly of toxic origin, the result of infectious diseases such as typhoid fever, influenza, pyæmia, gonorrhœa, erysipelas, syphilis, etc., and in part of a direct infectious nature.

They are divided according to their clinical course into acute (of a few weeks' duration), subacute (of some months'), and chronic forms.

The most frequent seat of the myelitic area is the dorsal cord and the most frequent form is of luetic origin.

Course.—The symptoms are very variable according to the seat of the lesion (see Section IV., 3).

The most common form (dorsal myelitis) causes a spastic paraplegia of the legs, without severe attacks of pain, although the latter may occur; disturbances of the bladder and of sensibility, paræsthesias; disturbance of co-ordination of the legs may also occur.

In lumbar myelitis: flaccid paraplegia of the legs, loss of reflexes, paralysis of the bladder, etc.

In cervical myelitis: spastic paraplegia of the legs, atrophy of the arm muscles, etc.

Bed-sores and cystitis frequently indicate the beginning of the end.

Diagnosis.—To be excluded are compression myelitis (which see). For localization of lesion see the same.

Treatment.—See compression myelitis.

When of syphilitic origin, mercurial inunctions.

The results of treatment depend upon the amount of degeneration that has gone on and the possibility of checking the process.

5. *Syringomyelia.*

(Plate 70, 1 and 3.)

Nature.—One includes under this name such focal diseases as lead to abnormal cavity formation within the spinal cord. Such cavity formations are dependent on many different causes. They may be congenital or they may develop later in life, a late congenital protal manifestation (hydromyelia), or they may arise from the occurrence of central glia proliferation (central gliosis), or from unknown causes after trauma, hemorrhages, etc.

The cavity formation may extend throughout the entire length of the cord and it may also extend into the medulla. The most frequent location is in the cervical cord, and therefore the most important symptoms are referable to this part of the cord.

As the result of cavity formation, which usually begins in the vicinity of the central canal, there is a slowly progressive destruction of the central gray substance of the anterior and posterior horns, and of the nerve fibres and cells that are therein contained.

Course.—The individual symptoms of such a condition are not characteristic; their combination or association, however, are pathognomonic.

They are:

(1) Progressive muscular atrophy and paralysis, especially of the upper extremities, in some cases of one side, in others of both sides. The small muscles of the hand, the triceps, etc., are most frequently

affected. The atrophied muscles show fibrillary twitchings and reaction of degeneration (degenerative atrophy, disease of the peripheral neuron cells).

(2) Disturbances of sensibility in the arms, especially analgesia and thermo-anæsthesia with intactness of the sense of touch (dissociation of anæsthesia).

(3) Trophic disturbances in the fingers and joints, such as necrosis, purulent formation, subluxation, and thickening.

To this there may be added spastic paralysis of the legs or such bulbar symptoms as spinal accessory and hypoglossal paralysis. Cases having an atypical course are not uncommon.

The process is a slowly progressive one.

Diagnosis.—To be excluded is spinal muscular atrophy (sensibility normal). Atypical cases are difficult to recognize.

Treatment.—Symptomatic; electricity, baths, etc., are useless.

6. *Hemorrhages into the Canal of the Spinal Cord.*

(Plate 56, 8.)

These may be extramedullary, localized in and between the membranes (intrameningeal apoplexy), or they may take place into the substance of the spinal cord (hæmatomyelia).

These hemorrhages are mostly traumatic, the intrameningeal form simulating in its symptomatology spinal meningitis, only that the symptoms are associated with and follow trauma. The symptoms are irritation root-symptoms, paraplegia, and disturbance of the bladder and of sensibility.

The symptoms of hæmatomyelia form in brief the same combination as syringomyelia, only that they come on in an apoplectiform way after trauma and vary in severity and extent according to the seat and extent of the lesion. The exact diagnosis is to be made by a consideration of the above-mentioned principles (sce page 193).

7. Multiple Cerebro-Spinal Sclerosis (Sclerose en Plaques).
(Plates 71, 1 to 4; 57, 5.)

Nature.—The nature of this disease still remains a riddle. It consists in the occurrence of numerous, small, irregular scattered foci of disease (primarily encephalitic?) in the brain and spinal cord. Within each of these foci the medullary substance and the nerve fibres that pass through perish, while the axis cylinders persist and functionate for a long time; there may be also an incomplete interruption of fibres. The disease occurs most frequently in the young after acute infectious diseases (?).

Course.—The symptoms are slowly progressive.

At first there is some disturbance in performing certain movements, weakness and uncertainty of the arms and legs, tremor on attempting to grasp an object (intention tremor), and the gait becomes weak and uncertain (spastic-ataxic).

Aside from these there is nystagmus, scanning speech (discontinuous, stopping speech), slight bladder disturbance, optic atrophy (not always). The patellar reflex is increased. From time to time apoplectiform attacks occur. Sensibility is but little disturbed; there may be paræsthesiæ. From this typical form there may be every imaginable deviation, dependent upon the localization of the foci, so that these "formes frustes" of multiple sclerosis give abundant opportunity to confound it with a number of other diseases. Many of the above-mentioned symptoms may be absent. Sometimes the ataxia comes prominently into the foreground, sometimes the spastic paresis of the legs, and sometimes the bulbar symptoms.

The disease pursues a very chronic course; transitory improvements are frequent. Toward the end intelligence may fail.

Diagnosis.—To be ruled out are: neurasthenia, hysteria (and similar disease conditions), dementia paralytica (pupils, history). Other spinal-cord diseases, such as myelitis, spastic spinal paralysis, cannot always be excluded with certainty.

Treatment.—Gymnastics, baths, rubbing, symptomatic.

B. SYSTEM DISEASES.

Concerning the nature of these see Section IV., 3, page 110. The greater number of them develop very slowly and are progressive and incurable. When the cells and fibres of the spinal cord undergo degeneration they cannot be restituted.

8. Spastic Spinal Paralysis.
(Plate 75, 1 to 4; Plate 57, 6.)

Nature.—This is a rare disease. It consists of a symmetrical primary degeneration of the pyramids (central motorial neuron complex), which goes on to destruction.

Course.—The disease begins in youth (children?) with slow and increasing spastic paralysis of the legs, arms, and muscles of the face, without muscular atrophy and without disturbances of the bladder or of sensation. The tendon reflexes are increased (patellar, foot phenomena).

Diagnosis.—There are other forms of double-sided spastic paralysis such as the cerebral (?) spastic paralysis of children, of multiple sclerosis, hydrocephalus, myelitis, etc. In these cases there will be accompaniment of other symptoms (mixed cases), but nevertheless not in all of them.

9. *Amyotrophic Lateral Sclerosis.*
(Plate 72, 2.)

Nature.—In this disease, in addition to disease of the pyramidal tract, there is involvement of the motor cells of the anterior horns (peripheral neuron), especially in the cervical cord and later of the medulla nuclei (see Section IV., 3).

Course.—The disease begins in adults with the appearance of weakness and atrophy of the muscles of the hand (interossei, thenar) and at the same time or perhaps a little later with a spastic paretic disturbance of gait. The atrophy in the arms develops slowly (for further details see "Spinal muscular atrophy"). Fibrillary twitchings and reaction of degeneration occur in the degenerating muscles and eventually the arm and the shoulder muscles entirely disappear, the reflexes remaining, even exaggerated, and the legs manifest the spastic paralysis. The patellar reflex is exaggerated; ankle clonus is present. Vesical and cutaneous sensibility are normal. Pain occurs especially in the arm muscles. To these symptoms may be added early in the course of the disease symptoms of progressive bulbar palsy (which see).

Treatment.—Symptomatic, electricity, baths, arsenic, feeding by means of the stomach tube.

10. *Spinal Progressive Muscular Atrophy.*
(Plates 72, 3; 56, 6; 55, 2; 77, 4; 58, 2 and 3.)

Nature.—A slow, progressive degeneration due to involvement of the motor ganglion cells in the gray anterior horns of the spinal cord, especially in the cervical portion. The pyramidal tracts are not involved. The muscular changes consist of a slow degenerative atrophy of the individual fibrillæ. There is a striking proliferation in the nuclei of the sheaths of the fibrillæ, the protoplasm diminishes, yet very small fibres may show distinct transverse striation. Toward the end this disappears like the rest of the protoplasm and there remains only the connective-tissue portion of the fibre.

Course.—The clinical symptoms are those of amyotrophic lateral sclerosis minus the spastic phenomena. The disease is very gradual in its development, the atrophy showing itself first in the small muscles of the hand, but eventually all the arm muscles are included. The shoulder muscles show fibrillary contractions and reaction of degeneration. The function of the bladder and sensibility are undisturbed. Duration from one to several years. The atrophy of the thenar muscles causes inability to bring the thumb in opposition with the other fingers, ape hand. The deltoid muscles atrophy frequently before the other upper arm and shoulder muscles.

The neck muscles, the intercostals, and very rarely the diaphragm may be involved.

Symptoms of bulbar palsy may follow the development of the disease or they may be simultaneous.

Diagnosis.—See below under Progressive muscular dystrophy.

Treatment.—See Amyotrophic lateral sclerosis.

Supplementary to this should be considered what is called neural muscular atrophy, the so-called peroneal type of progressive muscular atrophy (Plates 58, 4; 77, 6).

In this family form of atrophy the peroneal muscles, the tibialis anticus, etc., become diseased simultaneously or before the small muscles of the hand. This gives rise to a paralytic club-foot, following which is the peroneal gait; abnormally high elevation of the thigh, the tip of the foot being dragged. The patellar reflex is lost. Slight sensory disturbance of the toes and fingers. The muscular atrophy in the arms and legs progresses slowly. Anatomically there are found degenerative changes in the peroneal, median, and ulnar nerves, and also in the lower lumbar and cervical cord (anterior horn cells?).

11. *Progressive Muscular Dystrophy* (*Primary Myopathy*).
(Plate 58, 3.)

Nature.—It is impossible to say with certainty whether this disease is of spinal origin or not, as the changes that have so far been found have been in the muscles themselves. As the terminal member of the large group of motorial system diseases its consideration belongs here.

The muscle fibrillæ undergo an uncommonly slow atrophy, not degenerative. The individual fibres become smaller and toward the end disappear. The transverse striation remains for a long time intact. There is a proliferation of the muscle nuclei, an accumulation of fat interstitially and hypertrophic fibres are common (diameter of fibres 150–200 μ). The disease frequently occurs in families and during youth. The muscles in which the atrophy occurs show a simple, non-degenerative atrophy, there are no fibrillary twitchings and no reaction of degeneration. The atrophy is slowly progressive and recovery never results.

Course.—According to the beginning of the atrophy several forms have been described, which, however, toward the termination of the disease fall in with the single type.

These forms are:

(a) *Pseudohypertrophy.*

It begins in early childhood with weakness of the back muscles and of the muscles of the lumbar and upper thigh region. The muscles themselves are frequently unaltered in their form, but on account of the fatty infiltration (lipomatosis) they appear hypertrophic. These children cannot rise from the ground without helping themselves with the hands, the gait is waddling, and later walking is entirely impossible. The spinal column may be carious (lordosis).

(b) *Infantile Form.*

This begins in children, frequently with atrophy and paralysis of the face muscles (expressionless face, facies myopathique), to which is added later atrophy of the muscles of the extremities of the next form.

(c) *The Juvenile Form.*

This form occurs during and after the age of puberty. According to its distribution certain types have been made. The shoulder-girdle type shows atrophy of the muscles in the following order: the pectorals, trapezius, latissimus dorsi, serratus anticus, triceps, and biceps.

The small muscles of the hands are unaffected. The deltoid, the sterno-cleido-mastoid (the reverse of spinal muscular atrophy!) also remain free. There is in addition a thigh-girdle type, involvement of the quadriceps, gluteal, and peroneal, the calf muscles being uninvolved.

Both these forms pass one into the other and often develop simultaneously.

Diagnosis.—In order to diagnose the muscular atrophies one must familiarize himself with the appearance of the normal configuration of the extremities, then it becomes possible to recognize at a glance atrophy of the lumbricales (sunken spaces between the peripheral phalanges), of the thenar and hypothenar eminences (abnormally hollow), of the pectoralis (deep hollow under the prominently appearing clavicle), of the deltoid (flattening of the shoulder), of the trapezius (lessening in size of the neck), of the rhomboidei (abnormal sinking in in the place where normally there should be a pad), of the quadriceps (flattening of the anterior surface of the thigh), of the peronei (flattening of the normal pad going down from the cristæ tibiæ), etc. The diagnosis of any individual form is very simple if one observes carefully the beginning of the disease.

The treatment is symptomatic, orthopedic.

12. Poliomyelitis Anterior.

(Plate 56, 7.)

The inflammatory diseases of the anterior gray matter of the spinal cord occur in an acute or chronic form and injure especially the motor ganglion cells therein contained. They thus lead to degenerative muscular atrophy and to paralysis.

(a) Acute Poliomyelitis Anterior.

Nature.—It occurs in children in the early years of life (very seldom in adults). Its direct exciting cause we do not know. In form it is an inflammation of the substance of the gray matter of the anterior horns extending out from the blood-vessels. [It occurs also epidemically and then runs a course similar to the infectious diseases.]

Course.—In the beginning there is fever, vomiting, delirium. A severe comatose condition may exist for a few days, then convalescence sets in, but with this there is found a flaccid paralysis of the arms, legs (seldom the face), and bladder, with loss of reflexes and absence of sensory disturbances and pain. [Pain may be a very prominent symptom during the first few days of the disease.]

A part of the paralytic phenomena disappears after a time, another part continues (mostly one-sided). The muscles then undergo a rapid degenerative atrophy.

The residues of the disease which are met later in life are called "spinal paralysis of children." They are associated anatomically with a cicatricial shrinkage of the corresponding anterior horn (see Fig. 14, left anterior horn). Through the action of the antagonistic muscles there occurs contracture of the extremities. The affected extremity remains undeveloped.

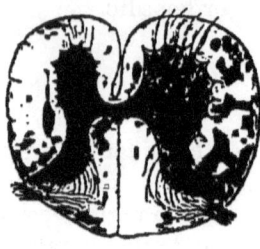

Fig. 14.

The most common forms of residue are:

Paralysis of the peroneal extensors (foot in position of pes equino-varus), paralysis of the upper-arm type (deltoid, biceps, brachialis, supinator longus), and paralysis of the lower-arm type (triceps and the extensors of the hand).

Treatment.—In the beginning, rest, baths, leeches over the spine, attention to the stomach, etc. Later, gymnastics and orthopedic treatment.

(*b*) *Chronic Poliomyelitis Anterior.*

Chronic atrophic spinal paralysis is a rare disease in adults which may cause a more or less rapid flaccid paralysis, first of the lower, then of the upper extremities, which is associated with degenerative atrophy of the muscles. The reflexes are lost. The bladder and sensibility are normal. Many muscles may remain unaffected. Recovery is possible.

To be excluded are lumbar myelitis (disturbances of sensibility) and multiple neuritis (which see).

Treatment.—Baths, electricity, symptomatic.

13. Tabes Dorsalis.

(Plates 72, 73, 77, 4; 57, 4.)

Nature.—Tabes dorsalis consists of a chronic progressive degenerative process in the peripheral sensory neuron complex, especially in those for the lower extremities. That portion especially is degenerated that comes from the spinal ganglia and passes into the cord; the long tracts in the posterior columns, the short ones in the posterior horns, Clarke's columns, and the fibres radiating to the anterior horns (see page 50). Degenerative processes also occur in the peripheral part of the neuron, in the sensory cutaneous nerves. In addition to these tracts there are often diseased other portions of the sensory (central) neuron complex, as well as a part of the motor peripheral and central neuron complex, thus constituting the combined form.

As a rule certain definite reflex tracts are first to perish (reflex collaterals?) and with these the patellar reflex and the pupillary reflex. There is still a great deal of obscurity as to the starting-point of the disease process.

For more explicit statements of the anatomy of tabes see Plates 73 and 74.

As has been said, the peripheral neuron complex of the lower extremities is most frequently affected, much more seldom that of the upper extremities (high tabes); isolated or combined with this form is affection of the peripheral neuron of the fifth nerve (especially the descending root) and of the glossopharyngeal. All these manifestations are parts of one and the same disease.

The cause of the degenerative process for the great proportion of the cases, if statistics are to be depended upon, is previous syphilis.

Course.—The symptoms, which develop slowly, may be considered in three stages:

1. *The Neuralgic Stage.*—There occur severe lightning-like pains which are very much worse at one time than at another. These pains radiate into the legs and are of a lancinating character (lancinating pains, irritation, posterior root symptoms). There are frequently paræsthesiæ, painful pressure sensations in the region of the stomach and abdomen, girdle sensation, and distention sensation.

If the patient now consults a physician there can almost always be found two symptoms which in combination bespeak pathognomonically tabes. They are:

(1) Reflex pupillary inactivity (loss of light reaction, preservation of the accommodative power, contracted pupils) [Argyll-Robertson pupil], and

(2) Loss of the patellar tendon reflex (Westphal's sign).

2. *The Ataxic Stage.*—This may occur after the first stage has continued for some years. It consists of the characteristic tabic, ataxic gait. The gait is flinging, stamping, uncertain, but not paretic. The uncertainty becomes greater when the eyes are closed and this constitutes Romberg's symptom (inability to stand with the feet together when the eyes are closed). In addition to this there are slight disturb-

ance of the bladder, lessened pain and temperature sensibility. Atrophy of the optic nerves may also show itself during this stage.

3. *Paralytic Stage.*—This occurs after a very variable period. The ataxia increases, is very distinct even when the patient is in bed, the gait becomes more and more paretic-ataxic, and at last impossible. Ordinarily no special paralysis occurs.

The bladder and sensory disturbances increase.

The previously mentioned lancinating pains which have been present increase and may be transformed to tabic "crises," during which there may be severe attacks of vomiting, coughing, difficulty in breathing, gastric, laryngeal, etc., crises. [It should not be forgotten that crises may be a very early symptom, preataxic in fact.]

Exudations into the articulations, arthropathies (trophic?), ulcer formation (mal perforant) are common.

Not infrequently earlier in the course of the disease there develop ocular palsies, paralysis of the trigeminus, and migrainous conditions.

Diagnosis.—The diagnosis is an easy matter in the ordinary case, but it may be rendered difficult by an uncommon onset, absence of important symptoms (no pupil immobility, preservation of patellar reflex, etc.).

To be excluded are multiple neuritis (which see) and hereditary ataxia (see also).

Treatment.—Anti-luetic treatment is of doubtful utility, treatment by baths, rest, electricity, massage, increasing nutrition, suspension, silver nitrate, ergot, narcotics, etc.

14. Hereditary Ataxia (*Friedreich's Disease*).

(See Plate 76.)

Nature.—This is a combined system disease; in addition to the degeneration of the posterior columns there is also degeneration of the lateral cerebellar tracts and of the pyramidal tracts. Involvement of the cerebellum is probable. The disease occurs in families.

Course.—The disease begins during youth with intense increasing disturbance of co-ordination in the movements; and maintenance of position of the extremities, the body, and the head is difficult.

Ataxia of the legs, arms, and Romberg's symptom occur early. Ataxia of the trunk is manifest when the patient is sitting as well as when he is moving. The gait is tumbling, simulating that of cerebellar ataxia, and paresis of the body and extremities gradually becomes more prominent.

In addition to these there are nystagmus, some disturbance of speech, and mild disturbance of sensibility and of the bladder.

Diagnosis.—The patellar reflex is lost, the pupillary light reflex normal (in contradistinction to tabes). The neuralgic pains characteristic of tabes are absent.

The disease progresses slowly until there is complete paralysis. Its duration is of many years. Recovery is unknown.

Treatment, symptomatic.

Similar to hereditary ataxia, there occur other forms of combined system diseases, which give rise in great measure to symptoms of spastic spinal paralysis combined with slight disturbances of sensibility and of the bladder. In this category belong the hereditary form of spastic spinal paralysis (involvement of the pyramidal tracts, the posterior columns, and the direct cerebellar tracts), and other forms which are less well known.

IV. DISEASES OF THE PERIPHERAL NERVES.

A. Diseases of Individual Nerves.

The most varying causes may lead to lesion of the peripheral nerves. Aside from trauma it may arise from tumors of all kinds (tumors of the soft parts, small exostoses in the bony canals, neuromata), cicatricial processes, inflammatory diseases which injure the nerves mechanically (compression neuritis); quite as frequently it arises from causes operating chemically, the formation of toxic substances during and after acute infectious diseases which injure individual nerves (toxic neuritis). It may also develop spontaneously (inflammatory). Disease of individual nerves occurs not infrequently (especially facial neuritis).

If the neuritic process is severe there is a complete degenerative destruction of the nerve fibres which eventuates in connective-tissue sclerosis. In the milder forms of the degenerative process (parenchymatous neuritis) complete restitution of the nerves may result (in contradistinction to lesion of the cord). The forms of the disease are classified according to their severity:

1. Diseases of the Motor Nerves.

All the above lesions when of purely motor or mixed nerves are attended by symptoms which are partly paralytic and partly irritative, such as localized muscular contractions, in the area of distribution of the nerve.

The paralyses that result are classified according to their severity (depending upon the severity of the neuritic process) into a severe, a moderate, and a mild form. (For further details concerning the diagnosis see Electro-diagnosis, IV., p. 116).

The most common individual paralyses are of the seventh nerve (due to rheumatism and diseases of the ear); paralysis of the musculo-spiral (most often compression palsy, sleep palsy, and toxic such as lead palsy); paralysis of the recurrent laryngeal (from aneurisms and tumors); paralysis of the spinal accessory; paralysis of the brachial plexus (birth palsies); and paralysis of the median and ulnar nerves (the result of wounds). For further elucidation of the symptoms of these and other possible paralyses see IV., 3, pages 102, and the following.

The most common localized muscular spasms from disease of individual peripheral nerves are:

Facial spasm, clonic contractions of the entire face musculature on one side, tic convulsif. Blepharospasm constitutes a partial form of this and consists of a tonic and clonic spasm of a lid.

Spasm of the muscles of mastication (trismus, a tonic variety of spasm frequently of central origin).

Spasm of the spinal accessory, spastic torticollis, a severe form of tonic-clonic spasm in the outer part of the trapezius, the sterno-cleido-mastoid, and often also in the other neck muscles innervated by the cervical plexus such as the splenius, etc. The latter may be involved alone.

Clonic spasm of the diaphragm, singultus, and spasm of the calves (crampus sensu strictiori), etc., may occur.

The cause of these spasms is nevertheless not always a lesion of the peripheral motor nerves, frequently they are produced reflexly, *e.g.*, by severe neuralgic pains.

It is often very difficult to differentiate between spasms of central (psychical) origin and those of peripheral causation. Both factors may be operative simultaneously to produce spasms; peripherally operative factors in centrally predisposed dispositions.

As forms of spasms occurring in this way may be reckoned the spasms of gaping, crying, laughing, yelling (see hysteria), saltatory reflex convulsions (contractions on attempting to walk), paramyoclonus multiplex (see hysteria), as well as many of the so-called occupation neuroses, which are frequently disturbance of the function of muscles used in the performance of certain work of a co-ordinative nature, such as writing, sewing, violin and piano playing, occurring in persons affected with neurasthenia.

Writer's cramp (mogigraphia) consists of involuntary cramp in the muscles of the hand and fingers associated with tremor, weakness, and painful sensitiveness of the hand. It may cause complete inability to write (spastic, paralytic, tremulous, neuralgic form). Psychical excitation and fear increase the inability (as in stuttering).

Treatment.—Occasional giving up of writing, stout penholder, Nussbaum's holder, cold applications, massage, gymnastics, electricity, use of the typewriter.

Remark.—Similar to the occupation spasms there occur occupation palsies (which may be associated with atrophy of the muscles) in connection with severe labor, such as in smiths (involvement of the small muscles of the hand), drummers (thumb muscles), and in milkers (ulnar paralysis), etc.

Treatment of the other palsies and spasms is scarcely necessary in the lighter forms; they recover by themselves, still electricity and massage may be used. In the severer forms treatment is of very little use. Electricity, massage, gymnastics, bathing, counter-irritation of the skin, cauterization in spasms, may be tried. For the psychical treatment see Neurasthenia and Hysteria.

2. The same variety of causes produces in the sensory nerves, anæsthesias in a similar distribution to the paralyses, and often severe neuralgic pains in areas corresponding to the irritative symptoms.

The last are of much more significance than the first. These periodically occurring pains, neuralgias, are frequently associated with paræsthesias and mild disturbances of sensibility in the area of skin to which the involved nerve is distributed. They follow most frequently in the wake of the infectious diseases, such as malaria and influenza, also after toxic diseases such as syphilis, diabetes, gout, and nephritis. They are produced mechanically by tumors, scars, neuroma; and lastly the spontaneously occurring forms (neuritis?) may be accompanied by skin diseases such as herpes zoster, urticaria, and erythema.

Diagnosis.—Neuralgia should be diagnosed only when the attacks of pain are confined to the exact distribution of the involved nerve and its branches, and furthermore when the nerve is sensitive, especially at the point of exit from its bony canal. Frequently it is possible to produce an attack of pain by pressure on such a point (points douloureux).

The most common neuralgias are:

Trigeminal neuralgia (tic douloureux, reflex facial spasm). In most cases only one branch is involved, thus causing supraorbital, infraorbital, inframaxillary neuralgia, and neuralgia of the tongue; pressure at the point of exit of these nerves (the supraorbital fissure, infraorbital foramen, mental foramen) produces pain.

Occipital neuralgia (occipitalis major), intercostal neuralgia (intercostal nerves), sciatic neuralgia (sciatic nerve), sciatica.

This most stubborn, frequent form of neuralgia causes pain extending from the hip to the knee and eventually to the outer side of the foot (distribution of the sciatic). Sensitive points are at the ischiatic foramen, the posterior surface of the upper third of the thigh in the middle of the gluteal fold, and in the middle of the popliteal space. Pain occurs when the thigh is flexed if at the same time the leg is extended (puts the nerve on the stretch). The leg muscles may atrophy and scoliosis may delevop. (To be excluded in making a diagnosis are carcinoma of the rectum, coxitis, tabes dorsalis. Examine the urine!)

Other forms of neuralgia are spermatic neuralgia (irritable testicle), coccygodynia, and articular neuralgia.

Treatment.—Discover the cause (syphilis!), galvanization (application of anodal weak current), counter-irritation of the skin, mustard leaf, blister plaster, scarification, sweat baths, hot baths, warm applications better than cold massage.

Use may be made of such anti-neuralgics as the salicylates, antipyrin, analgene, etc., and of arsenic and potassium iodide.

Only in most severe cases should morphine or cocaine be used. As last resources resection of the nerve and stretching may be tried.

B. MULTIPLE NEURITIS.
(Plate 78.)

Nature.—This consists of the simultaneous disease of several nerves and is almost entirely of toxic or infectious origin. It occurs after diphtheria, tuberculosis, typhoid fever, influenza, erysipelas; it also results from syphilis, gonorrhœa, diabetes, alcoholism, lead and arsenic poisoning, etc.

Anatomically the disease consists for the most part of a parenchymatous degeneration (primary destruction) of the medullated nerve fibres, accompanied by secondary interstitial changes (round-cell infiltration, sclerosis). Perineuritis and interstitial neuritis occur less frequently than this parenchymatous neuritis. Involvement of ganglion cells in the anterior horn is very probable in many cases (and not always secondary), but the significance of such involvement is disputed, *i.e.*, whether primary or secondary.

Course.—Multiple neuritis comes on rather acutely with general febrile symptoms, increased and painful sensibility, and paræsthesia which is soon followed by appearances of paralysis of muscles, which in severe cases is accompanied with degenerative atrophy. Its distribution is mostly symmetrical. In some cases there may be ataxic disturbances of motility combined with muscular paresis. The muscles most commonly involved are the extensors of the arm and legs, the psoas and the peroneal; but any muscle may be involved.

The reflexes within the involved area are lost; there are such disturbances of sensibility as hyperæsthesia, paræsthesia, and anæsthesia. The muscles may manifest all the various forms of reaction of degeneration (see IV., p. 120). The nerve trunks and muscles are painfully sensitive to pressure.

After a variable time, from weeks to months (acute, subacute, chronic forms, see Electro-diagnosis), there follows slow recovery. Entire recovery is the rule, but severe cases may terminate fatally in a short time from involvement of the vagus or from complications.

Depending upon whether the sensory or motor nerves are more involved, there has been separated a motor, a sensory, and an ataxic form, but a strict limitation to any of these forms is not possible, the symptoms of a single case being made up of these three varieties together. The ataxic form, on account of its similarity to tabes, has been called pseudo-tabes. The principal forms of multiple neuritis are:

(a) *Alcoholic Neuritis.*

Motor form: flaccid, double-sided paralysis of the peronei, the quadriceps, the glutei, the common extensors of the fingers and toes, the extensor pollicis, etc.

The patellar reflex is lost and there occur disturbances of sensibility and paræsthesiæ.

In other cases the prominent symptoms are radiating pains, paræsthesiæ, sensitiveness to pressure of the diseased nerves and muscles, while in others ataxic disturbances of the gait with muscular paresis and later atrophy (the pupillary light reflex preserved, contrary to tabes) are the symptoms that come prominently into the foreground. Not infrequently there exist at the same time other manifestations of alcoholism, such psychical disturbances as delirium, maniacal state, and mental confusion.

(b) *Diphtheritic Neuritis*

occurs generally as a motor form and especially of the muscles of deglutition (accessory vagus), of the eye muscles (diplopia), and of the muscles of accommodation. In severe cases the patellar reflex is lost and there may be a flaccid paralysis of the extremities and vagus paralysis.

(c) *Lead Neuritis.*

(In type-setters, painters, etc.) This form involves especially the muscles of the posterior surface of the forearms, hands and fingers which are supplied by the musculo-spiral nerve, the supinator longus escaping. In addition to this there may be manifestations of sensory irritation such as lead colic, paræsthesiæ, and mild disturbances of sensation.

(d) *Arsenical Neuritis.*

It produces especially severe sensory disturbances (paræsthesia of the fingers, pains, disturbances of sensibility) in addition to the symptoms of paralysis.

Remark.—The name acroparæsthesia has been given to a painful paræsthesia of the finger tips, without motor disturbances, a disease of unknown origin somewhat similar to arsenical neuritis.

[It is more than probable that acroparæsthesia is most often directly dependent upon and associated with vasomotor and vascular variations.]

(e) *Infectious Polyneuritis.*

This is of unknown, but certainly of infectious origin. The disease begins with general febrile symptoms, severe neuralgic pains which are spontaneous and producible by pressure. Rapidly developing muscular weakness leads to flaccid paralyses, mostly of the legs, then later of the arms (ascending variety). The tendon reflexes are lost, and disturbances of sensibility, œdema, tachycardia, difficulty of respiration (involvement of vagus) occur later.

The duration of the disease is very variable. Recovery is the rule, but death may occur from involvement of the vagus.

This form has heretofore been known as Landry's paralysis. Changes in the spinal cord may be present in the form of certain (secondary) degenerations. There occur also cases with accompanying foci of myelitis (myelo-neuritis). Their exact diagnosis is difficult and the prognosis is unfavorable.

Remark.—A very similar clinical picture is caused by acute poliomyositis, a disease etiologically related to infectious polyneuritis. In these cases the sensitiveness of the muscles precedes the neuritic pain for several days. There are furthermore no disturbances of sensibility, no reaction of degeneration, but there are general febrile symptoms, severe paralysis, and œdema. Trichinosis is to be carefully excluded.

Anatomically there are found inflammatory changes in the muscles (round-cell infiltration, etc.).

Myositic changes are also frequently found in cases of primary neuritis.

Diagnosis.—In order to establish the diagnosis of neuritis in its different forms it is necessary to exclude tabes (immobility of the pupils, disturbance of the function of the bladder, ocular palsies, prominence of the ataxia, absence of muscular paresis in the beginning of the disease, the history of the patient); lumbar myelitis (bladder disturbance, no sensitiveness of the nerve trunks to pressure); poliomyelitis anterior (no disturbances of sensibility, no sensitiveness to pressure, no spontaneous pain). Examination of the urine for sugar, albumin, and lead is of the greatest importance.

Treatment.—Removal of the cause (alcohol, lead), possibly anti-luetic treatment, rest, salicylates, antipyrin, and later the use of electricity, gymnastics, massage, and baths.

V. OTHER DISEASES OF THE NERVOUS SYSTEM IN PART OF UNKNOWN NATURE AND SEAT.

1. *Basedow's disease* is a term applied to a symptom complex consisting of exophthalmos, struma (vascular), tachycardia (pulse 100 to 160) in combination with other nervous manifestations. These are tremor, abnormal sweating, heat feelings, and

irritability. The bulging eyeballs are not properly covered by the lids (Stellwag's symptom) and when the eyes are directed downward there is a dissociation between the movement of the globes and the movement of the lids (Graefe's symptom). The internal recti muscles are often insufficient (Möbius' symptom). The disease has been supposed to be dependent upon abnormal activity of the thyroid gland (toxic), while others consider the seat of the disease to be the medulla.

The symptoms may be of very varying severity and atypical cases are not very infrequent. The duration of the disease may be many years.

Treatment.—Bromides, arsenic, phosphorus, sodium carbonate (30 to 150 grains a day), attention to nutrition, electricity, local applications, baths. Digitalis is useless. [The careful use of aconitia, $\frac{1}{500}$ grain pills, 4 to 10 to be taken daily, is strongly recommended by Seguin.]

2. *Myxœdema.*

Myxœdema develops simultaneously with the disappearance of the thyroid gland and the disease stands in relationship with disease of this organ; loss of its secretion apparently causing the diseased process (opposite of Basedow's disease?). The disease has followed complete extirpation of the gland (cachexia strumipriva).

The symptoms of myxœdema are: the occurrence of a disfiguring œdema in the skin of the face and hands. The skin becomes thick, peeling, and dry. Movements become uncertain and clumsy. Then follow mental apathy and dementia.

Treatment.—Feeding with thyroid substance is followed by good results, as is injection of the juice of the thyroid, the glycerin extract, tablets of the thyroid and desiccated thyroids internally.

[The tablets of Burroughs, Wellcome & Co. have given best results in this country.]

3. *Acromegaly.*
(Plate 77, 3.)

In this obscure disease there occurs progressive enlargement of the fingers, hands, feet, nose, lips, jaws (the ends, "akra," of the body) which leads to permanent deformity. There may be atrophy of the optic nerves and ocular paralyses the result of pressure of an enlarged hypophysis in the sella turcica. Glycosuria is often present.

Other diseased conditions which might belong here have been briefly considered in Section IV., 4. They are hemiatrophia facialis, circumscribed cutaneous œdema, erythromelalgia, and symmetrical gangrene.

4. *Tetany.*

This consists in the occurrence of recurring attacks of spasms especially of the flexor muscles of the fingers (claw hand), arms, and toes. The spasms occur symmetrically and can be produced by pressure on the trunks of the nerves (Trousseau's phenomenon). The attacks may recur for several weeks and longer.

Remark.—Tetaniform spasms, especially of the flexors of the fingers and of the calf muscles, are very common symptoms of neuritis in alcoholic patients and in individuals who have some affection of the stomach. They may also occur in healthy individuals (hand-workers).

Tetanus is a disease which follows infection of a wound by the tetanus toxin of the tetanus bacillus. Through such secondary infection, even after slight wounds, there may arise tonic convulsions of the entire musculature of the body.

SECTION VI.

Remarks upon the Conduct of an Autopsy and upon the Methods of Procedure in the Microscopic Investigation of the Nervous System.

At the autopsy on cutting into the cavity of the brain and spinal cord one should be mindful from the beginning that an eventual later microscopical investigation of the central organs may have to be made. Such an investigation must frequently be carried out for a more exact establishment of anatomical relations, inasmuch as the macroscopic condition oftentimes gives information only of gross focal lesions but never of the various forms of degenerative diseases.

Unfortunately this guiding point of view is oftentimes ignored.

The autopsy in such cases should be made as quickly as possible after death, never later than twenty-four hours. In diseases of the spinal cord the corpse should be allowed to lie upon the belly until the autopsy is made, in order to avoid disturbing post-mortal blood gravitation.

The greatest error made in autopsies that are generally carried out by rule of thumb is due to thoughtless cutting, pressing, and tearing; by such procedures a thorough, systematic investigation is oftentimes made impossible.

Even in the removal of the brain sufficient pain is oftentimes not taken. If the membranes (tentorium, etc.) are not completely loosened, the cervical cord and cranial nerves not entirely severed, the parts must be torn forcibly apart and this may cause entire destruction, for example, of the soft parts of the brain.

In the next place many, for the most part unnecessary, incisions are made into the central ganglia, corpora quadrigemina, etc., which make later orientation in serial sections very difficult. It is customary to make incisions through the central ganglia obliquely, and thus such dislocation of masses of the soft brain occurs that in later microscopic section they are unrecognizable. An exactly horizontal or frontal section should be made, or better none at all on the fresh brain.

The brain should never come into contact with water, otherwise subsequent hardening becomes in part illusory.

It is best, after opening a diseased focus by horizontal, superficial incisions, not to insist upon a revelation of all details (not very much more is attained by ever so many incisions), but to undertake an exact investigation first after hardening in Müller's fluid* has been completed. This can be done after a few weeks without harm if the brain is kept at a warm temperature (30° C., not more) and if the precaution be taken to open the lateral ventricles and to sever the corpus callosum. In this way orientation

---

* Bichromate of potassium 2 parts, sulphate of soda 1 part, and distilled water 100 parts.

is much easier and a recent, previously invisible degeneration is recognizable at the first glance.

The spinal cord is often maltreated by the chisel and forceps. Diseased, softened portions particularly tear at every strong pull or pressure. Here also one should not in the beginning make too many incisions. Nothing can be made out with certainty in the fresh state. It is well in separating the posterior roots to carry the knife as far outward as possible; in this way one is able oftentimes to retain a part of the spinal ganglia with the posterior roots, especially in the lumbar cord.

In the removal of diseased nerves and muscles one should be guided somewhat by a consideration of the course which the disease had run; otherwise one will rue too late some omission. Nerves should be bound to small pieces of wood and, as with the muscles, labelled carefully (name, side, level of the removed part, and date).

The protocol of the autopsy should be as complete as possible. The condition of the bony cavities of the skull and the spinal cord should be observed, also the state of membranes and vessels. The contents of the ventricle should also be remarked before the removal of the brain.

The microscopic investigation can be carried out in some cases (in recent focal diseases) immediately after the autopsy without previous hardening of the central organ. A small part is cut off with the scissors and teased upon the slide. Granular cells, broken-down nerve fibres and cells, and broken-down muscle fibres may in this way be recognized.

They will be more distinctly apparent after a rapid

staining (by addition of a drop of one-per-cent methyl blue or of hæmatoxylin).

Degenerated parts may be put after removal and after they are cut into small pieces in a one-per-cent osmic-acid solution which should be kept in the dark. After twenty-four hours the destructive products (fat corpuscles) will be stained very black. The results will be better if pieces are treated after the method of Marchi, *i.e.*, place small pieces first for a short time in Müller's fluid (not alcohol!) and then for twenty-four hours in osmic acid.

For all systematic microscopical investigation the process of hardening needs to be carried out with the greatest care. This consists in the proper use of Müller's fluid and alcohol.

If the purpose be to investigate specially the nerve fibres for degeneration the parts should be hardened in Müller's fluid. If the ganglion cells are to be carefully studied for the results of inflammation then the hardening should be in alcohol.

Often it is necessary to carry out both of these procedures, in which case it is most serviceable to harden the great mass of the brain, cord, nerves, and muscle in Müller's fluid, while small parts are removed from various levels and prepared for the reception of the nuclear stain by being hardened in absolute alcohol. In those that are prepared in the latter way bacteriological investigations may be made.

The hardening in Müller's fluid requires for the brain several months, but the process may be very much shortened by keeping the solution at a temperature of 30° C. A higher temperature should be avoided.

The solution should be changed with sufficient frequency so that it remains clear. At first (for the first week), particularly if it be kept warm, it is necessary to change it daily, later very much less often. As the hardening progresses it is well from time to time to make new incisions (horizontal or frontal) into the substance of the tissue to facilitate the permeation of the chrome salts.

Examination should not be undertaken until the process of hardening is entirely completed; in this way misleading artificial products will be avoided.

Hardening in alcohol is only a matter of a few days and it is necessary to change the absolute alcohol frequently.

After the hardening in Müller's fluid is completed small pieces of the parts to be examined are put in alcohol without previous washing in water and thus the hardening is completed. After this, for the purpose of serial sections, comes the embedding in celloidin, first in a thin, then in a thicker solution. After from five to twenty days, longer for larger pieces, the pieces are fastened to corks and put in a seventy-per-cent solution of alcohol until cut, which may be done if it is desired after a few hours.

It is not necessary that the sections be cut very thin. If it is desired to make an uninterrupted serial section through the entire piece of tissue or cord, as the case may be, a number of slides are prepared by dropping on one slide sufficient of a thin solution of celloidin to cover it thinly. Then strips of toilet paper of the width of the slide are used to remove the sections from the knife of the microtome as fast as they are cut. In order to prevent the sections on the strips of toilet paper from drying while a sufficient series to cover a slide are being obtained, it is necessary to keep the strips moistened with alcohol. This is most easily done by saturating a piece of blotting paper in seventy-per-cent alcohol, which is kept on the plate on which the pieces of toilet paper rest while the series is being cut and while a number of series is being prepared preparatory to being transferred to the slides. This transference is accomplished by simply inverting the strips of toilet paper on to the side of the slide that has been prepared with celloidin.

Staining is then carried out in the parts hardened in Müller's fluid according to the method of the Weigert-Pal stain for the medullary sheaths and by the carmine stain for the ganglionic cells (eventually also double carmine staining) and in the pieces hardened in alcohol according to Nissl's method, the technique of which can be found in the text-books of Kahlden, Friedlander-Eberth, and others.

By means of the stain for the medullary sheaths one gets a black coloration of all medullated nerve fibres; in case where these are not properly preserved, that is, degenerated or disappeared, the black coloration does not take place, *i.e.*, such areas appear to remain unstained, light. The fat cells of the medullary sheaths make a chemical combination with the hæmatoxylin coloring matter, the expression of which is the black coloration.

The modern method of silver impregnation of Golgi and Cajal can be used for pathological elucidation only under certain conditions, inasmuch as it is irregular in its action.

A knowledge of the structure of the nervous system is to be reached by the following methods of investigation.

1. By the making of a series of serial sections throughout the normal developed organs of man and its subsequent reconstruction after they have been impregnated with the various pigments (Stilling's, Meynert's, etc., method).

2. By the investigation of pathological cases, especially by secondary degeneration (methods of Türck and Waller, etc.).

3. By the method of comparative anatomy and comparative embryology (method of Meynert and Edinger).

4. Through the developmental method (observation of the time of appearance of medullation) (Kölliker's, His', and Flechsig's methods).

5. Through investigation by means of artificially produced degenerations (Gudden's, etc., method).

These methods have been made use of as far as possible in the preparation of the plates of this atlas.

LIST OF THE ABBREVIATIONS USED IN THE ILLUSTRATIONS.

a, anterior.
Ab, basilar artery.
Acc, arteria corporis callosi.
al, ansa lentiformis (loop of the lenticular nucleus).
ad, descending root of the auricular nerve.
AS, Aqueduct of Sylvius.
AfS, Arteria fossæ Sylvii.
Av, vertebral artery.
av, vertebral arch.
B, brachium (especially brachium conjunctivum of the cerebellum, processus cerebelli ad corpora quadrigemina).
Bulb. olf, olfactory bulb.
c, conarium.
Ca, anterior horn (spinal cord).
ca, anterior commissure (spinal cord and brain).
ca, cornu anterior (lateral ventricle.)
cAm, cornu Ammonis.
Cb, cerebellum and lateral cerebellar tract in the spinal cord and medulla.
cc, corpus callosum.
cc, corpus candicans (mammillary body).
ce, external capsule.
cerv, cervical.
cgl, lateral geniculate body.
cgm, median geniculate body.
Ch, chiasm.
Ci, internal carotid.
cia, anterior crus of the internal capsule.

cip, posterior crus of the internal capsule.
cL, Luys' body (*c.sbth*, corpus subthalamicum).
Cl, Clarke's column.
Cl, claustrum (brain).
cm, median commissure.
Cp, posterior horn (spinal cord).
cp, posterior horn (ventricle).
cp, posterior commissure.
cqa, anterior quadrigeminal body.
cqp, posterior quadrigeminal body.
cr, restiform body.
c.sbth., corpus subthalamicum.
cst, corpus striatum.
ct, central tegmental tract.
ctr, corpus trapezoides.
cun, cuneus.
CV, centrum semiovale of Vieussens.
d, dura mater.
DBr, decussation of brachium conjunctivum.
Dec. Pyr., decussation of pyramids.
f, fornix.
fa, remains of anterior column.
fae, external arcuate fibres.
fai, internal arcuate fibres.
fal, remains of antero-lateral column.
fB, column of Burdach.
fc, calcarine fissure.
fd, descending fornix.
fG, column of Goll.
Fl, flocculus.
fl, lateral limiting layer.
fl, } posterior longitudinal fasciculus.
flp, }

231

ABBREVIATIONS IN THE ILLUSTRATIONS.

fp, posterior column.
F.rhomb, rhomboid fossa, fourth ventricle.
fs, fossa of Sylvius.
G, Gowers' columns.
g.ang, angular gyrus.
g.cent, central gyri.
g.fr, frontal gyrus.
g.H, gyrus hippocampi.
glp, globus pallidus (internal lenticular nucleus).
gl, lateral geniculate body.
gm, median geniculate body.
go, occipital gyrus.
Go, Gratiolet's visual radiation.
gsm, supramarginal gyrus.
gt, temporal gyrus.
K, raphe of the tegmentum.
L, Lissauer's field.
L₁ lenticular nucleus.
l, lateral.
L pc, paracentral lobule.
m, median.
M.ob, medulla oblongata.
n, nodulus.
N.am, nucleus amygdalæ.
NB, nucleus of Burdach.
nc, caudate nucleus.
nfa, nucleus arciformis.
NG, nucleus of Goll.
nVIIId, dorsal nucleus of auditory nerve.
nVIIIv, ventral nucleus of auditory nerve.
nr, red nucleus.
O, olive (inferior, large).
ol.m, median accessory olivary nucleus.
ol.p., posterior accessory olivary nucleus.
ol.s, superior olive.
p, pia mater.
p, posterior.
Ped, peduncle (cerebral).
Pl.ch, choroid plexus.
Pulv, pulvinar.
Put, putamen.
Py, pyramidal tract.

qa, anterior quadrigeminal body.
qp, posterior quadrigeminal body.
R, raphe of the tegmentum and pons.
ra, anterior root
Rc, ramus arteriosus communicans.
rp, posterior root.
R.subth, subthalamic region.
Sa, anterior longitudinal sulcus.
s, solitary fasciculus (descending root of the ninth and tenth).
Sg, substantia gelatinosa (posterior horn).
Sn, substantia nigra.
Sp, posterior longitudinal sulcus.
sp, septum pellucidum.
Srt, substantia reticularis tegmenti.
str.s. (*m* and *i*), superior (middle and inferior) layer.
st.a, striæ acusticæ.
t, tænia thalami.
T, teg, tegmentum.
Th, optic thalamus.
tr.o, optic tract.
trig. olf, trigonum olfactivum.
U, uvula (vermis).
unc, uncus (gyrus hippocampi).
V, ventricle (lateral, median, third).
vIV., fourth ventricle.
Vs, superior vermis of cerebellum.
Vi, inferior vermis of cerebellum.
v, vertebra.
vma, velum medull. ant. (with lingula).
zr, root zone.
V'c, caudal, descending root of trigeminus.
V'm, motor nucleus of trigeminus.
V'n, nasal, ascending root of trigeminus.
V's, sensory nucleus of trigeminus.
VIIIcN, cochlear branch of auditory nerve.
VIIIvN, vestibular branch of auditory nerve.

www.ingramcontent.com/pod-product-compliance
Lightning Source LLC
Chambersburg PA
CBHW051726300426
44115CB00007B/486